COMPUTER CONCEPTS
AND MANAGEMENT
INFORMATION SYSTEMS

COMPUTER CONCEPTS AND MANAGEMENT INFORMATION SYSTEMS

C. P. Gupta, PhD

&

K.K. Goyal, PhD

MERCURY LEARNING AND INFORMATION
Dulles, Virginia
Boston, Massachusetts
New Delhi

Publisher: David Pallai
MERCURY LEARNING AND INFORMATION
22841 Quicksilver Drive
Dulles, VA 20166
info@merclearning.com
www.merclearning.com
1-800-232-0223

C. P. Gupta & K. K. Goyal. *Computer Concepts and Management Information Systems.*
ISBN: 978-1-68392-586-6

The publisher recognizes and respects all marks used by companies, manufacturers,
and developers as a means to distinguish their products. All brand names and product
names mentioned in this book are trademarks or service marks of their respective
companies. Any omission or misuse (of any kind) of service marks or trademarks,
etc. is not an attempt to infringe on the property of others.

Microsoft Corporation screenshots fall under the guidelines seen here:
https://www.microsoft.com/en-us/legal/intellectualproperty/permissions/default.aspx .

Library of Congress Control Number: 2020936528

202122321   Printed on acid-free paper in the United States of America.

Our titles are available for adoption, license, or bulk purchase by institutions,
corporations, etc. For additional information, please contact the Customer
Service Dept. at 800-232-0223(toll free).

All of our titles are available in digital format at *www.academiccourseware.com* and other
digital vendors. The sole obligation of MERCURY LEARNING AND INFORMATION to the purchaser
is to replace the book, based on defective materials or faulty workmanship, but not based on
the operation or functionality of the product.

CONTENTS

PREFACE

In today's world of computer advancements, no professional training or course in higher education is considered complete unless readers are exposed to the fundamentals of information technology, including hands-on experience with *Microsoft Office* and a knowledge of information systems. This book covers all of the major concepts of these subjects which are actually related to one another. Chapters in the book are organized in the following manner: Chapter One introduces the concept of the computer system, its evolution, input-output and storage devices (also termed as the "hardware component" of a system). Also, it clears very basic concepts such as improving system processing speed, etc. Chapter Two introduces the reader to software and the graphical user interface (GUI). Chapter Three of the book covers four of the most important and popular software applications, *MS Word, MS Excel, MS PowerPoint* and *MS Access*. Chapter Four and Chapter Five introduce the reader to management information systems, ERP, CRM, their utility for the creation of an enterprise, and topics such as security, business ethics, and cybercrimes. We are certain this book will help readers in having a better understanding of these topics. Creative suggestions for improvement of the book would be acknowledged gratefully.

ACKNOWLEDGMENTS

Any endeavor is incomplete unless acknowledgment is given to all those who are behind the curtain. We are immensely thankful to the higher authorities of our institution that always motivated us to excel in our field by providing facilities and the highest level of support. We are thankful to our director, Dr. B. B. S. Parihar, without whose push this book could not have seen a place on the shelves. We are thankful to all our colleagues who supported us in our efforts. Any acknowledgment is incomplete if it does not include the family. We are indebted to our parents, spouses, and children who left us in our domain to complete the work on this book and never complained for not getting enough attention and time from us.

C. P. Gupta
K.K. Goyal
April 2020

CHAPTER 1

CONCEPTS AND COMPUTER FUNDAMENTALS

1.1   Introduction

The extraordinary development of information technology during the past few decades has changed the way business is done in the real world. In the early years of IT development, very few big corporate houses were capable of buying computers and using them because they were very costly and their implementation required a unlimited technical and professional skills and aptitude to operate them. Nowadays the situation is reversed: computer systems, which were costly, have become affordable even for a middle-class family. Large computer systems have been reduced in size and only require a small amount of desktop space.

Previously, computer programming and operations required hard work, time, and professional help to learn, but today, thanks to the invention of the graphical user interface (GUI), users can easily operate any software. With the introduction of wireless technology, the utility of a computer system has been further enhanced as wiring requirements have been significantly reduced and with the availability of the laptop, it is possible to carry a system from one place to another. During transit it is very easy for the user to communicate with their clients and complete pending work.

Introduction of low-cost computer systems and a lower-cost Internet has helped in the prevalence of computer systems in a wider section of society. In the present era, information technology has the potential to influence the lives of ordinary citizens as much as it influences business, education, and government. The high penetration of smartphones, computers, and the Internet, is changing the lives of people across the globe. The information superhighway (bringing IT, entertainment, and the communication industry onto one platform), which was a concept in the nineties is now a reality because of the availability of smartphones. Today, a PC is no more considered a device to be used by only IT professionals, rather it's considered essential to a household.

1.2   Data, Processing, and Information

Data: Data can be defined as *a collection of raw facts and figures which in itself has no meaning.* Take for example, "10, 15." The 10 and 15 can be two numbers, two weights, two distances, the height of any two persons or they can be the measurement of any two liquids. Thus, one can say that data has no meaning unless and until it has been given a shape of some meaningful result.

Processing: Processing is done on raw data which gives it a meaningful form. In a computer, four types of processing activities can be performed. They are:

Calculation	This includes addition, subtraction, multiplication, and division
Comparison	This includes comparisons like >, >=, <, <=, <> etc.
Decision Making	This includes making decisions on a basis of a condition
Logical Branching	This means, based on the decision made, jumping from one part of the computer program to another

Information: Information can be defined as the processed data that has meaning.

For example: Add the numbers 10 and 15.

In this example the data is the numbers 10 and 15. Processing is the addition of these two numbers; and the information is 25, which is the result of this addition.

Can We Say Data and Information Are Interchangeable?

Yes, data and information are interchangeable. What is considered information in one instance may be considered data in another. For example, the grades of all the students in a class in all subject areas are *data*. When these grades are calculated to find out a percentage, the percentage marks are the *information*. But when a teacher is willing to find out the percentage result of his class, all the individual percentage marks of all the students is called *data*. When the teacher adds these percentage marks of all the students and divides them by the total number of students, the obtained result is *information*, such as a class percentage.

Why There Is a Need for Data and Information

In today's world, business organizations are facing cut-throat competition in the marketplace. It has become very difficult to survive and to secure the relevant market share; it is becoming harder to *maintain* the market and the market share. Only proper access to data and the information generated from that data can help business organizations make quick and relevant decisions. These decisions not only help business organizations in retaining their market shares, but they also help in keeping track of their competitors' activities in the marketplace.

In today's business organizations, data and information are no longer treated as mere tools for conducting business, rather they are considered important assets, which help them in making proper and timely decisions at various levels of management. For example, decision support systems at the middle level of management, and executive support systems at the top levels of management which processe data to generate information.

1.3   Defining The Computer System

A *computer system* can be defined as an electronic device and thus has two states, one when current flows in circuits, represented as "ON" and second, when current *does not* flow in circuits, represented as "OFF." These two states in a computer system are represented by a binary number system which consists of two digits: "1" and "0." The "ON" state is represented by "1" and the "OFF" state is represented by "0." A computer system performs four tasks for the user as follows:

- It accepts data.
- It stores the data.
- It does the processing.
- It gives the output or result to the user.

A computer system accepts data in the form of alpha numeric characters for example 2A/127 Govind Nagar Agra-282004, or in numeric digits for example 100, 250, 6285 or in alphabetic form, for example "Robert Smith."

A computer system can also be called an information processing system because it:

- manages voluminous data perfectly,

- provides confirmation of the validity of data and transaction.

- performs the complex processing of data and multidimensional analysis,

- helps in quick search and retrieval of related data,

- provides mass storage,

- provides timely information to the user, and

- it is adaptable, as per changing needs of individual users and corporations.

A railway reservation system is a perfect example of this, as it handles millions of reservations daily, and checks for the data entered such as the correct train name, date, gender of customer, etc. It processes thousands of reservations across the country simultaneously, searches the status of millions of tickets in virtually no time, and provides the latest updates to travelers, thus providing all the information required by users.

1.4   Computer Classification

A computer system can be classified into the following types:

On Purpose Basis: On the basis of *purpose*, computers can be classified as:

- **General Purpose Computers:** Computers that perform regular work such as data analysis, accounting, generating bills and receivables, billing payables, stock management, etc. Computers used in offices for commercial, educational, and other applications are included in this category.

- **Special Purpose Computers:** Computers that perform special tasks such as weather forecasting, space applications, medical diagnostics, etc.

On Technology Basis: On the basis of the *technology*, a computer system can be grouped into three categories:

- **Analog Computers:** Analog computers are those computers that measure quantities such as current, voltage, frequency, pressure, temperature, speed, etc.; and convert them into their numeric equivalent. For example:

  - A thermometer that measures the rise in mercury level and converts it into its numeric equivalent.

  - Machine gasoline pump that measures the flow of liquid and converts it into its numeric equivalent.

- **Digital Computers:** Digital computers are those computers in which all the processing is done in binary digits (0's and 1's).

- **Hybrid Computers:** Hybrid computers are those computers which process analog signals and convert them into digital signals and vice-versa. Hybrid computers are mainly used in artificial intelligence (robotics) and computer-aided manufacturing (e.g., process control).

On the Basis of Memory Size and Capacity: According to the size and *memory/storage capacity*, there are four types of computers:

- **Microcomputer:** Microcomputers are also known as desktop PCs or personal computers and serve a single user at any given time. They are also known as "stand-alone systems" and consist of a main chip called a microprocessor. A microprocessor is a chip that consists of an arithmetic and logic unit (ALU) and a control unit (CU). A company called INTEL makes microprocessor chips. The extended technology of "(XT)" PCs have various versions of the microprocessors with names like 8086, 8087, 8088. The advanced technology of "(AT)" PCs include versions 80486, P1, P2, P3, and P4. The P# series is known as the Pentium series. As of today, the line-up of INTEL Core processors include the Intel Core i9, Intel Core i7, Intel Core i5, and Intel Core i3, along with the Y-Series Intel Core CPUs.

- **Minicomputers:** Minicomputers serve multiple users at the same time and are general-purpose systems. They have more processing power and are more expensive than the microcomputers. Unlike microcomputers, minicomputers have a single central processing unit (CPU) and have various terminals attached. A terminal consists of a monitor, keyboard, mouse, and sometimes a printer. For example, see the IBM 9375, PDP-1.

■ **Mainframe:** Mainframe systems can support thousands of users at a time. They are similar to minicomputers but with greater storage and processing capabilities. Identifying numbers of terminals supported by a mainframe are much higher in comparison to the minicomputers. For example, the IBM system/370, IBM 4300 series are mainframe systems.

■ **Supercomputers:** Supercomputers are designed to process complex scientific applications and are the most powerful and the most expensive computers. They are based on the principle of parallel processing which is also known as a "Non_Von Neumann Design." In parallel processing there is one main processor and to it are attached various coprocessors and all work simultaneously. Main usages of supercomputers are in the field of climate forecasting, petroleum exploration, nuclear energy research, defense, etc. For example, CRAY-3, CRAY-2, ETA-10, PARAM are supercomputers.

On the basis of the generations of a computer: On the basis of *development*, a computer can be classified into the following generations:

■ **First Generation (1940-1956):** First-generation computers were built before the 1960s. These computers used thermionic valves or vacuum tubes for the purpose of making circuits. These computers were not reliable as they consumed large amounts of electricity, and the vacuum tubes that were used in these computers generated a enormous amounts of heat causing frequent breakdowns. They used magnetic drums for memory. Some examples of first-generation computers are the UNIVAC and the ENIAC.

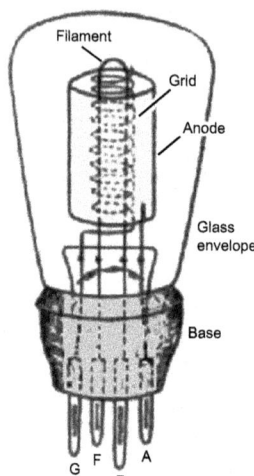

Thermionic Valve/Vacuum Tube

- **Second Generation (1956-1963):** In the second generation of computers, thermionic valve/vacuum tubes were replaced by transistor technology. The development of the transistor technology helped in the development of a smaller, faster, and more reliable computer system. This development also resulted in the improved efficiency and storage capacity of a computer system. It was this development that made the computer system more popular and reduced the prices. Assembly language replaced the binary language during this period.

Transistor

- **Third Generation (1964-1971):** Individual components were interconnected in the first and second generation of computers to form the circuits, but with the introduction of integration technology it became possible to have more than one circuit packed into a single integrated circuit container known as an "IC Chip." This development reduced the size of the computer significantly and increased the data storage and processing capabilities to an enormous level. Punched cards were replaced by keyboards in the third-generation computer as an input device.

Integrated Circuit Chip

- **Fourth Generation (1971-present):** Development of *large-scale integration* (LSI), and *very large-scale integration* (VLSI), further reduced the size of computers and increased the processing speed and storage capabilities. This development made it possible to have thousands of integrated circuits built onto a single silicon chip.

VLSI- Chip

- **Fifth Generation (Artificial Intelligence):** Development of *ultra large-scale* integration (ULSI), led to the dramatic reduction in the size of computers, and increased the processing capabilities of a system beyond imagination. With this technological development, computers were now capable of supporting a very large storage-capacity hard-disk, optic disk, multimedia, and Internet capability, etc. Parallel processing is now helping to make artificial intelligence a reality.

ULSI- Chip

1.5   Block Diagram

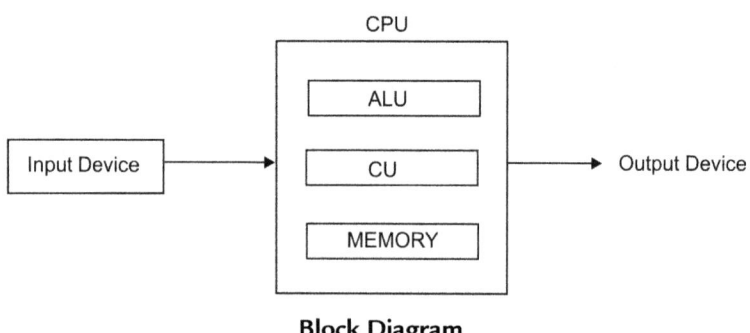

Block Diagram

Data in a computer system is entered with the help of an input device. Once the data reaches the central processing unit, the control unit directs the data into the memory. This means as soon as data is entered in the computer system, it first gets stored in the random access memory (RAM), or primary memory of the system. When a command is given to a system for processing, the data from the memory is transferred to the arithmetic and logical unit for processing. Processing generates the result, which is again directed by the control unit to the memory of a system. Once an instruction is issued to produce the output, this result from the memory is directed to an output device by the control unit. Let us understand this with an example.

```
10 Let A = 15
20 Let B = 25
30 Let C = A + B
40 Print C
50 End
```

As soon as the variables A and B are assigned a value of 15 and 25, a memory location with a name A and B will be opened in the RAM of the system with values of 15 and 25 stored in it. The next instruction is C = A + B. Now the data from the RAM will be transferred to the ALU for processing and the result (which is 15 + 25 = 40) will be stored in another location "C" in the RAM of the system. The next instruction is "print C," the control unit will read the value of the location C in the RAM and will display that on the monitor of the computer system. The command "End" will tell the computer that program is over.

Various devices of a computer system form a BLOCK DIAGRAM:

- **Input Device:** The input devices are used to enter data and instructions into a computer system. These devices act as a linking point for the external environment of a computer system to its internal environment. These input devices accept the data in English or any other language from the user and then convert the data entered by the user into the machine code, which a computer can understand. The most commonly used input device is the keyboard.

- **Central Processing Unit:** The central processing unit (CPU) is also termed the brain of a computer system. As in humans, the brain controls all of the activities; similarly in a computer system, it is the CPU that controls all of the processing functions. It has the following main parts:

  - **Arithmetic and Logic Unit:** The arithmetic and logic unit (ALU) is a place where all the functions are performed. The ALU not only executes the mathematical calculations, but it also performs the logical comparison and decision making. Logical comparison and decision making are the functions that make a computer system different from a calculator.

– **Control Unit:** The control unit (CU) acts as a supervisor of the system. It is the responsibility of the CU to synchronize and coordinate all the activities performed by a computer system. The CU acts as a traffic policeman and directs the transfer of data from one part of the CPU to another and vice versa.

■ **Memory:** Memory, which stays inside a central processing unit, is known as a primary memory. It is in the form of a silicon chip in which data is stored in the form of electronic pulses. The presence of current is shown as "1" and absence of current is shown as "0." Data in this memory is stored in the form of 0's and 1's.

■ **Output Device:** After processing the data, the result is generated and it is directed to a device called an *"output device."* This device may be a monitor (visual display unit) or a printer attached to a computer system, or a hard disk, an optical disk, etc.

1.6 Hardware Versus Software

All the physical components of a computer system that a user can touch are termed as the *hardware* of a computer system, for example, the keyboard, the monitor, CPU, mouse, the printer, etc. However, *software* is that part of a computer system that we cannot touch and we can only see, such as an Internet browser, Microsoft Office, etc. Software can be defined as the program that instructs a computer how to process the data and generate required output.

1.7 Input Devices

Input devices are used to enter data into a computer system. Much development has taken place in input devices. In the first generation of computers, expert knowledge was required to punch data into a computer system, but now input devices give great ease to the user while they enter data into a system. The various kinds of input devices used with a computer system evolved as follows:

■ **Punched Card:** In a punched card, information is punched as holes. They consist of 80 columns and each column consists of 12 positions, which can be punched. They are inexpensive, but expert knowledge is required to work with punched cards and it is very difficult for a common user to maintain and control them.

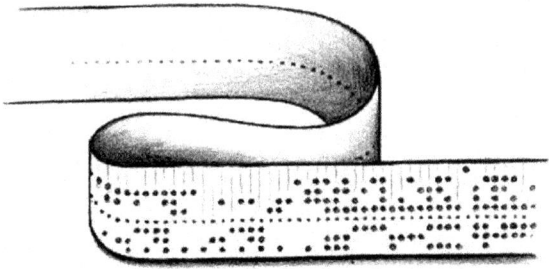

- **Paper Tape:** Paper tape and punched card work on the same concept. The difference is that a paper tape is a continuous strip of paper, whereas a punched card is in the form of a small card. Characters are formed in a paper tape using a code, which consists of circular holes made across the width of the tape.

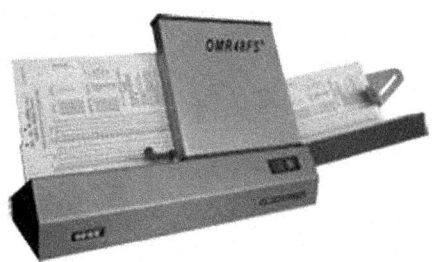

- **Optical Mark Reader:** Every competitive examination makes use of an *optical mark reader* (OMR) sheet. In OMR, marks (in the form of an oval or a circle) are made with the help of a pencil or a pen. Evaluation is done by throwing a light on the OMR sheet and the reflected pattern is matched with the correct pattern, which is already available in the system.

■ **Optical Character Reader:** OMRs were only able to detect the presence or the absence of the marks, and this drawback of OMR was removed with the development of the *optical character reader* (OCR). An OCR is able to identify any character. OCRs read each character with the help of a photoelectric device that determines the outline and shape. The shape is read and then compared with a shape that is already stored in the system. OCR is advantageous as the sheet can be read directly by a reader, and the input goes straight into the computer system for processing.

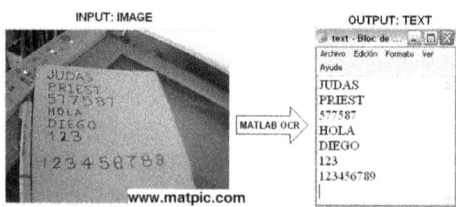

■ **Magnetic Ink Character Recognitions (MICR):** MICR is used mainly by banks. Using this method, the documents with characters marked with ink are passed through a magnetic field where the ink-coded characters magnetize the reader's head due to the magnetic ink. MICR helps in the faster processing of the instrument.

■ **Bar Code:** A bar code consists of a series of black bars and white spaces in between those black bars. The bars are of varying widths, and they are printed on labels to uniquely identify items. The bar code labels are read with a scanner, which measures reflected light and interprets the code into numbers and letters that are passed onto a computer. These codes are specific codes, some of the more common being the Universal Product Code (UPC) and European Article Number (EAN). Bar codes are especially used in labeling numerous consumer products and books.

- **Keyboards:** Keyboards resemble a typewriter and are the most widely used input devices. They consist of keys that represent digits, alphabets, and special symbols. They also have function keys from F1 to F12, the use of which varies depending on the software being used. Most companies use a keyboard with 104 keys.

- **Mouse:** The drawback of earlier input devices (movement restriction) were removed with the introduction of the *mouse*. With the use of GUI, there arises a need for an input device that can help with the inputting of data by selecting an option on the desktop. With the help of a mouse it became possible for a computer user to have a 360-degree rotation facility on the screen, which was otherwise not possible. A mouse comes in two varieties: One includes a tracking ball that generates the signal to move a pointer on the screen, and the other is an optical mouse that senses the movement and moves the pointer on the screen.

- **Speech Synthesizer:** In a speech synthesizer data is entered in a system in the form of a human voice with the help of a microphone. The system converts this data into electronic signals. These signals are then matched with patterns that are already available in a computer system. One problem faced by the speech synthesizer is that if one changes the way one speaks, the computer may not recognize the pattern of the input voice.

- **Scanner:** Scanners are handheld devices, and are used to scan complicated diagrams, pictures, and graphics that are otherwise difficult to draw with the help of other input devices. Flatbed scanners are now easily available and are very easy to operate and can also scan large pictures. Nowadays, printers with inbuilt scanners are also available. An example of a scanner is the HP M1005 all-in-one printer.

- **Light Pen:** A light pen has a photocell at its tip. It is moved on the screen and to touch the required option. The light pen senses the light coming out from that option and executes the file behind that option. The light pen is mainly used for the graphical work and in computer-aided designing (CAD). Light pens are widely used during football telecasts, during which the commentators draw free-hand lines on the TV screen.

- **Touch Screen:** Invention of the touch screen has been considered a revolution in the field of input devices. Touch screens are extensively used in smartphones, ATMs, railway enquiry systems, and many other places. They are very easy to operate. Users just need to touch the option they want to select. Touching the option breaks the light beam emitted, and thus the position of the option is recorded and the program behind that option is executed.

1.8   Output Devices

These devices show the *output* of the processing of a program. A computer can generate two kinds of outputs, one known as *soft copy* and is in the form of a computer file, which can be either stored in any storage device or displayed on the monitor of a system. The second is in the form of a *hard copy*, also known as a computer printout. Various types of output devices are as follows:

- **Printers:** Printers are used to produce hard copy of output and are divided into two categories, one is an *impact printer* and second is a *nonimpact printer*:

    - **Impact Printers:** Impact printers work similar to a typewriter. In a typewriter, characters are formed on the paper when an arm with a character embossed strikes the ribbon and forms the impression; similarly in an impact printer there is a head that consists of a number of pins (usually nine or twenty-four) that strikes the ribbon, which in turn forms the impression of a character on the paper. Impact printers are also called *dot matrix* printers. Dot matrix printers can be divided into following categories:

- **Character Printers** are printers that print single characters one at a time from left to right, and then from right to left.

- **Line Printers** are printers that print a complete line at a time from left to right, and then from right to left.

- **Page Printers** are printers that print a full page at a time.

■ **Nonimpact Printers:** These printers never touch the paper. They form the image of a character on the paper with the help of heat or a laser. Nonimpact printers can be divided into the following categories:

• **Thermal Printers:** They work on the concept of heat. The papers, which are sensitive to heat, are used and characters are formed in dotted form. Some drawbacks of this printer are that a special kind of a paper is required, and they cannot print multiple copies at a time.

• **Laser Printers:** Laser technology is used by these printers for the purpose of printing. The laser beam charges the drum on which the ink powder (called toner) is thrown and gets deposited on the characters formed on the drum. When the paper rotates on the drum, these characters are printed on the paper. The initial cost of a laser printer is high, but the per page printing cost is comparatively low.

• **Inkjet Printer:** Inkjet printers are low in cost in comparison to the laser printer, but the cost of printing is more. This printer uses the electric field and throws the ink from the nozzles on the paper. The paper absorbs the ink and forms the characters.

- **Plotters:** Plotters consist of an arm that can rotate 360 degrees and can print. Plotters are used mainly for the printing of technical designs used in computer-aided designing (CAD) or in computer-aided manufacturing (CAM).

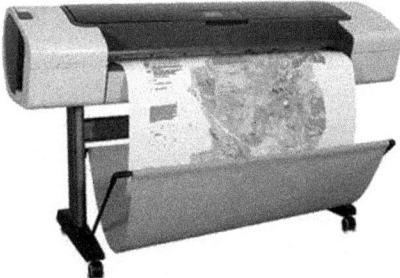

- **Monitor (Visual Display Unit):** A monitor resembles a TV screen and is used for showing output.

1.9 Computer Memory

Memory is a place where the data and the instructions are stored in a computer system. The memory of a computer system can be divided into two categories: *primary memory* and *secondary memory*.

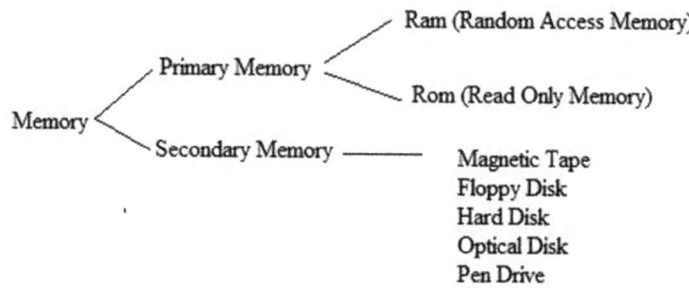

Primary Memory: Primary memory is the memory that is found inside a computer system. Data in the primary memory is stored in the form of electronic charges, and that is why this memory is temporary in nature. The moment the computer is switched off, data written in primary memory will be wiped out. The measurement unit of the memory is bits and bytes and can be defined as follows:

A Bit = 0 or 1

A Byte = any combination of 8 bits.

1024 Bytes = 1 Kilo Byte (KB)

1024 Kilo Bytes = 1 Mega Byte (MB)

1024 Mega Bytes = 1 Giga Byte (GB)

1024 Giga Bytes = 1 Tera Byte (TB)

In a computer system one character needs one byte of memory space for the purpose of storage. For example, if one wants to store "RED" in a computer system, then one needs three bytes of memory space. Primary memory can be further divided into two categories.

ROM: ROM stands for *"Read Only Memory,"* and as the name suggests in ROM we can only read. We can neither write nor erase whatever is written in a ROM chip. A ROM chip is required to execute instructions, which are very frequently executed by a computer system. Because of this, these programs and instructions cannot be stored in RAM as it is volatile in nature; hence these are permanently stored in a ROM chip and are placed inside the CPU. The program, which is usually stored in a ROM chip, is the part of the operating system called the *basic input-output system* (BIOS program). It starts as soon as the computer is switched on and makes the computer ready to load the rest of the operating system in the memory of a computer so it is ready to work.

ROM has a few variants as follows:

- **PROM** is known as *programmable read-only memory* in which data can be written once and then it cannot be altered. PROM is sold as empty, and can then be filled with a program by the user. Once filled with the program, the contents of PROM cannot be removed.

- **EPROM** is known as *erasable programmable read-only memory* in which data can be rewritten many times, and for this purpose the EPROM chip has to be removed from the CPU and exposed to ultraviolet rays so that new data can be written onto it.

- **EEPROM** is known as *electrically erasable programmable read-only memory* that can be reprogrammed using special electronic pulses a number of times without removing it from the CPU.

RAM stands for *random access memory*, and as the name suggests in RAM we can read, we can write, and we can erase whatever is written into it. RAM is also known as read/write memory because data can be read from a ram chip and can also be written onto it. It is a volatile memory, and as soon as the computer is switched off the data written in the RAM is wiped out. Various types of RAM on a PC are:

- **DRAM (*Dynamic RAM*):** It needs to be refreshed periodically by the CPU so that the data contained in them is not lost.

- **SRAM (*Static RAM*):** In it data contained remains stored properly; therefore it does not need to be refreshed by the CPU. This type of RAM has a higher speed than DRAM and is costly, too.

- **EDO RAM (*Extended Data-out RAM*):** It is basically used in the Pentium systems and is suitable for having bus with speeds up to 66 MHz.

- **SDRAM (*Synchronous Dynamic RAM*):** It can be considered as an extension of DRAM, but has a higher speed than DRAM. It is suitable for a system bus with speeds up to 100 MHz.

- **RDRAM (*Rambus Dynamic RAM*):** is a type of memory that is faster and more expensive than SDRAM. This memory is used on systems that use the Pentium 4.

Secondary Memory: *Secondary memory* is a permanent memory and remains outside a computer system. In this memory data is stored in the form of magnetic particles on hard disk, floppy disk, magnetic tape, and in the form of pits on the optical disk.

1.10 Storage Devices: DASD/ SASD (Direct Access Storage Devices/Sequential Access Storage Devices)

Storage devices are referred to by a variety of names such as auxiliary storage, auxiliary memory, secondary storage, and backup storage. They are capable of storing large amounts of data. They are used as an online extension to the main memory; and are also used for offline storage of programs and data.

Differences between DASD and SASD

The workings of *direct access storage devices* (DASD) and *sequential access storage devices* (SASD) can be understood as follows:

DASD: These devices allow the user to access any record directly. Examples of these types of devices are floppy, CD, DVD, hard disk, etc.

SASD: These devices do not allow the user to access any record directly. For example, if a user needs to read and record the number 50, they have to first bypass record 49, only after that they would be able to read the desired record. It works like a cassette of a tape recorder. An example of this type of device is *magnetic tape*.

Various storage devices are described in the following sections.

1.10.1 Magnetic Tape

Magnetic tape was used as offline storage for large amounts of data because it is inexpensive.

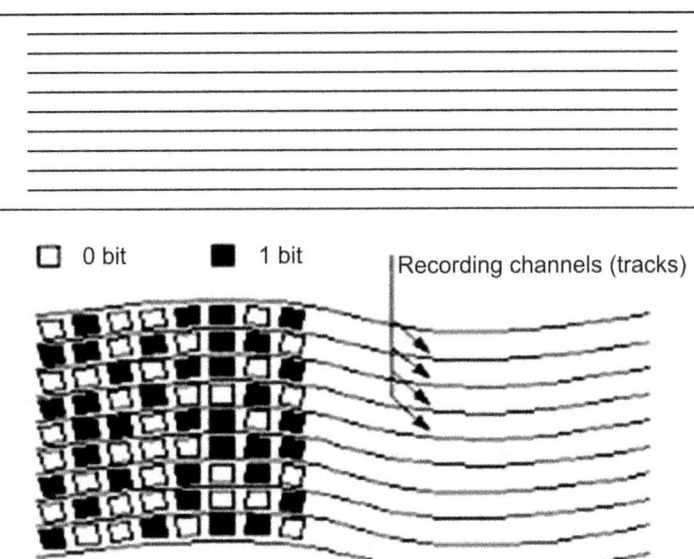

A magnetic tape is similar to an audio tape. A metal foil called a marker indicates the beginning of the tape. Data is stored one character at a time; either 7 or 9 bits format is used for each character and they are recorded in parallel across the width of the tape. Data on the tape is saved in the form of records separated by a gap called *inter record gap* (IRG). The tape always remains in motion and is only stopped when a record gap comes below the read/write heads. The tape motion is stopped only when the read data is to be transferred to the memory of the system. During the time the tape is transferring the data into the memory any further reading process is not completed. So the IRG is given to allow the tape to attain its normal speed before the beginning of next record is reached. A group of related records is called a file, and a file marker identifies the beginning of the file. It is a specially coded record preceded by a gap longer than the record gap. The first record following the file mark may be used as a header or identifier for this file, and the last record may be used as a trailer or end for this file.

*		Record		Record		Record		*
*		d	IRG	d	IRG	d		*
*		1		2		3		*

In the nine tracks format a set of nine heads are mounted to read/write information on tape. Each head operates independently and stores information along one track of the tape. While eight tracks are used to record a byte of data, the ninth track is used to record a parity bit for each byte. The parity bit checks if the data has been read/written accurately or not. The recording density is measured in bits per inch, that is, bits per track of the tape. For example on one inch of a 9-track tape having a recording density of 1600 bits per inch the total number of bits stored is = 1600*9 = 14,400 bits. The tape travels at a speed of 100 inches per second, and during the time the tape takes to accelerate to its full speed, no recording takes place.

If the record size is of lesser length than the block fixed for it, the rest of the block is left blank. Because of this a lot of space is wasted. To reduce this gap on tape, records can be blocked together and in place of *inter-record gap* (IGR) we will have *inter-block gap* (IGB). For example, in a blocking factor of 3, three records per unit are recorded.

*		Record		Record			*
*		1, 2, 3	IBG	4, 5, 6	IBG		*
*							*

1.10.2 Floppy Disk

Mylar plastic coated with magnetic oxide is used for making a floppy. This flexible material is cut into circular pieces of 5.25 inches or 3.5 inches in diameter. Because of the flexible material used during production, they are called *floppy disks*. These were small, low in cost, and could be very conveniently carried from one place to another. In a floppy disk, data is stored in the form of magnetic particles on the tracks. A hub in the center is used for mounting the disk into the disk drive. Because there is a long slit in a floppy provided for the read/write head to access the data, there are many chances of the disk becoming unusable because of its exposure to dust, scratches, etc. Data could easily be stored and retrieved with the help of the floppy disk. The floppy disk had a longer life in comparison to magnetic tape, but the data is less secure.

For a standard IBM formatted double-sided, high-density 5.25 inch floppy diskette, the following properties applied:

- Data is recorded on two sides of the disk

- Single-sided, 9 sectors/track: 180 KB

- Double-sided: 360 KB

- High-Density (HD): 1.2 MB

For a standard IBM formatted double-sided, high-density 3.25-inch floppy diskette, the following properties applied:

- Data is recorded on two sides of the disk

- Each side has 80 tracks

- Each track has 18 sectors

- Each sector holds 512 bytes (0.5 KB)

- Each floppy disk holds 2880 sectors (2 * 80 * 18), for a total of 1440 KB or 1.44 MB

1.10.3 Hard Disk

Magnetic disks or the "hard disk" or Winchester disk were first introduced in 1956 for the purpose of bulk data storage. A hard disk contains circular platters that are made of any metal or aluminum and are coated with magnetizable material. The number of platters depends on the disk capacity. The higher the number of platters, the higher the data-storage capacity of the disk. To be able to store data on the disk, it is essential to format it first. Formatting the disk creates magnetic tracks and sectors where data is stored in the form of magnetic particles. A conducting coil called the *drive head* is used to store/retrieve data from the disk. When a user tries to read/write data onto the disk, the head remains stationary while the platter rotates beneath it. Data is stored on both sides of the disk on concentric rings called *tracks*. Each track is of the same width as that of the head. To minimize errors due to the interference of the magnetic field, adjacent tracks are separated by gaps called *intratrack gaps*.

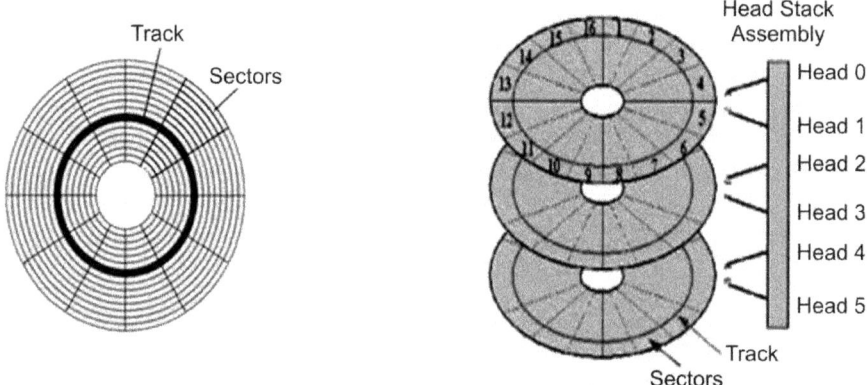

Data is stored and read from the disk in blocks called *sectors*. The heads are mounted on a rigid arm that can be extended or retracted across all the tracks and are either fixed or movable. The speed of disk rotation is 7,200 revolutions per minute. Nowadays, disks can rotate at a speed of 10,000 revolutions per minute. The higher the speed of the rotation of the disk, the higher the rate of the transfer of data into the computer memory. This increases the chances of the disk becoming heated which will result in a shorter life of the hard disk.

All the platters rotate at a constant speed around the spindle. The drive head, while positioned close to the center of the disk, reads from a surface that is passing by more slowly than the surface at the outer edges of the disk. Because of this, tracks toward the outer side of the disk are less densely populated with data in comparison to the tracks are toward the center of the disk. This results in the reading of same amount of data over the same period of time, either from the outermost track or from the innermost track.

Position of the read/write head is at a fixed distance from the platter allowing an air gap. The read/write process always starts at the sector boundaries. A sector can store a maximum of 512 bytes. If the lesser number of bytes are to be stored in a sector, the rest of the sector is padded (filled) with the last byte recorded. For the purpose of storage, in a hard disk clusters are always allotted. A cluster consists of a number of sectors that are always an exponent of 2. The only odd number of sectors a cluster could consist of is 1 ($2^0=1$). For example, if there is a file with a size of 1,000 bytes, one cluster (two sectors) would be allocated to the file on a disk, later, if data is appended to the file and its size grows to 1.600 bytes, another two clusters are allocated, storing the entire file within four clusters. If adjacent clusters are not available, the second two clusters may be written elsewhere on the same disk or within the same cylinder or on a different cylinder—wherever the file system finds two sectors available. A set of corresponding tracks on all surfaces of the disk pack equidistant from the spindle is called a *cylinder*.

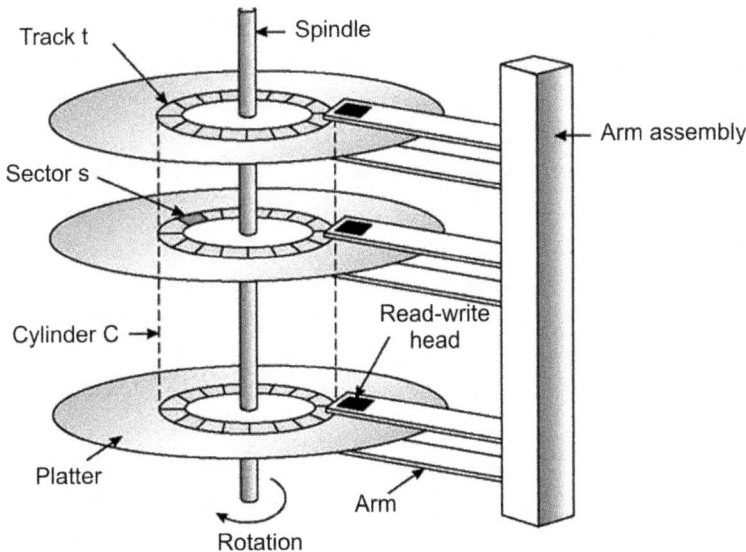

The hard disk allows the direct access of records. That is why they are also known as the *direct access storage devices* (DASD). To read/write data, the cylinder number, the surface number and the sector count must be known.

1.10.4 Optical Disk

The latest developments in hardware and enhanced processing speed of systems generated a need for quick and fast storage. Data is stored in the form of the magnetic particles in earlier data storage devices, but with the development of optical technology it is now possible to store data in the form of tiny particles called pits (created by a laser beam). Data is stored on the optical disk in the form of light particles which do not generate a magnetic field, thus they can be stored very close to one another. In optical disks, streams of digital data in the form of tiny pits are burned onto a thin coating of metal or other material deposited on a disk. A beam of laser light is used to read these pit patterns. When it encounters a pit, the intensity of the reflected light of the laser changes. The change is detected by photo sensors and converted into a digital signal. The disk can store up to 600 MB of data and can be addressed by track and sectors.

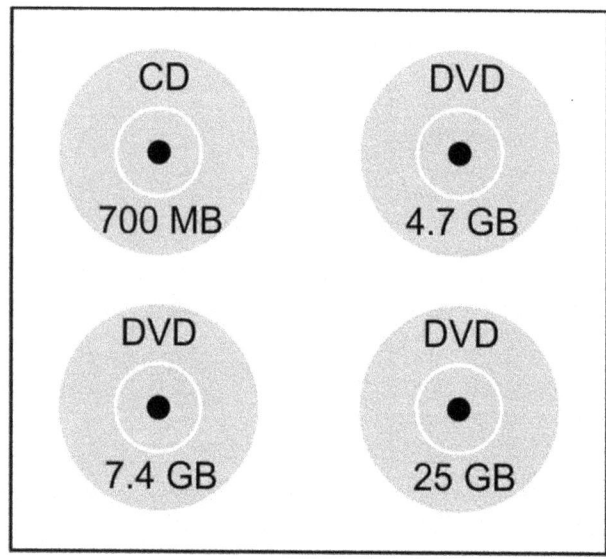

Optical disks are available in the following forms:

- **CD/CD-ROM (*Compact Disc* or CD-Read-Only Memory)**: This is a very commonly used term and usually refers to non-writable discs.

- **CD-R (*Writable Discs* or CD Recordable)**: This means that a user can only burn (store) data a single time on it, or multiple times when using multisession mode until it reaches the disk capacity. Once data is written it cannot be overwritten or erased.

- **CD-RW (*Rewritable* Discs/CD Rewritable)**. This means that a user cannot only write data, but can also erase the data written on the disc and can write new data on it.

- **DVD/DVD-ROM (*Digital Versatile* Disc/DVD Read-Only Memory)**. Basically the same as CD-ROM, however, a DVD typically has a six times greater capacity than a CD.

- **DVD-R/DVD+R (*Writable DVD* or DVD Recordable)**. Basically the same as CD-R with greater capacity.

- **DVD-RW/DVD+RW (*Rewritable* DVD/DVD Rewritable)**. Basically the same as CD-RW with greater capacity.

- **DVD DL (*Double* or Dual Layer)**. Double-layer discs have twice as much disc space as standard DVDs.

- **BD-R (*Blu-Ray* Disc Recordable)**. It is a Blu-ray disc on which data can be written only once.

- **BD-RE (*Blu-ray Disc Rewritable*)**. It is a Blu-ray disc recordable erasable (BD-RE), which can be recorded on and erased any number of times.

Blu-ray is a digital optical-disc data-storage format that has a data storage pattern that supersedes the DVD format. Blu-ray discs are capable of storing hours of video in high-definition and ultra high-definition resolution.

1.10.5 Pen Drive

A pen drive is a portable storage drive that can be carried easily from one place to another and is a very popular storage device among users (popularly called PD by users). A pen drive is very easy to operate; the user simply needs to put it into the Universal Serial Bus (USB) port and it is ready to use. A pen drive works on the principle of EEPROM.

1.10.6 Flash Memory

Flash memory can be defined as a nonvolatile memory that can be erased and reprogrammed in units of memory called blocks. It works on the concept of an EEPROM and the only difference is that in EEPROM data is erased at a byte level, whereas in flash memory it is done at a block level making flash memory faster. Flash memory is often used to hold control code such as the basic input/output system (BIOS) in a personal computer. When BIOS needs to be changed (rewritten), the flash memory can be written to in block (rather than byte) sizes, making it easy to update. However, flash memory is not as useful as random access memory (RAM) because RAM needs to be addressable at the byte level and not at the block level. Flash memory gets its name because the microchip is organized so that a section of memory cells are erased in a single action or a "flash." Flash memory is used in digital cellular phones, digital cameras, LAN switches, PC cards for notebook computers, digital set-up boxes, embedded controllers, and other devices.

1.11 Basic System Configuration

A *Basic system configuration* is something that all users should understand before buying or working on a computer system. If not properly configured, a system will give problems to its users while working. It means working out the details of all the parts as per the user's requirements that need to be assembled to complete a computer system. Types of parts used to assemble a system greatly affect the working of a computer. Basic system configuration encompasses full details of main memory (RAM), hard disk drive, card reader, microprocessor, DVD drive, monitor, networking requirements and details of an operating system. A user should not randomly buy things offered on the market. Instead they need to research their requirements. If a user needs the system for software development, then the requirements would be different. However, if the user needs the system only for general purpose work like working on the Internet or using MS Office, then the requirement of system configuration would be on the lower side. Hardware like the monitor, keyboard, mouse, and speakers should be configured wisely for better performance. Properly configuring a computer requires a lot of thinking, and the user needs to know what they are actually taking home might affect the performance of the system. The answer to a question such as "What is the configuration of your computer?" would be based on the points that follow:

1. What is the type of processor (CPU) do you have?

2. How much memory (RAM) is in the system?

3. What is the capacity of the hard-disk drives (HDD)?

Other details like the size and type of monitor, keyboard, mouse, etc., can be defined to complete the description of the system configuration. A user can find the configuration of its system by following these steps:

- Click on the "System and Security" icon in the control panel and under it, select the "System" option. The user can right-click on the "My Computer" icon on the computer's desktop to find information about which version of Windows (Operating System) is on the computer, what kind of processor (CPU) it has, and how much of memory (RAM) is installed.

- Another way of obtaining the details of the configuration of the computer system is by first clicking on the "Start" option that appears on the status bar of Windows and then typing "msinfo32" in "search programs and files" option.

- Information about the hard-disk capacity can be obtained by double-clicking on the "My Computer" icon on the desktop. A window will be opened and will provide the details of all the hard-disk partitions and their storage capacity.

1.12   Processing Speed

There are mainly two factors that affect the performance of a computer system; one is the number of bits processed at a time, and the second is the clock speed of a computer system. The number of bits processed at one time depends on the bus and is explained as follows:

Computer Buses: A bus is used to carry people from one place to another. In a computer, a *bus* is a special circuit that connects the various parts of the computer and allows the transfer of data between these parts. There are various kinds of computer buses operating in a PC. They are:

- *Address Bus*: A set of address wires that give the memory address used by the data, and is referred to as an address bus.

- *Data Bus*: A set of wires that transfers the input/output data from and to the memory is known as the data bus.

- *Control Bus*: A set of wires that controls the read/write operations is known as the control bus.

The processing speed of a computer is controlled by the number of bits it can process at a time. If the data bus contains 8 wires, it can process 1 byte or 8 bits at a time and is called an 8-bit processor. If there are 16 wires then the data bus can process 2 bytes or 16 bits at a time and is called a 16-bit processor. There are 32-bit and 64-bit processors with an ultra-high speed. Thus one can say that one factor that affects the performance of a PC is the number of bits processed at a time.

Clock Speed: It is defined as a speed with which data is transferred inside the CPU from one place to another. It is measured in mega or gigahertz. Higher the clock speed, the faster will be the processing of a system.

1.13   Uses of a Computer System

Computers are extensively used in any or every field now a day. A few of the prominent areas where a computer system has significant uses are:

- **Education:** Educational institutions across the world cannot even think of operating without using a computer system. Whether it is a library or a classroom setting, or an office, computers are extensively used at all levels in an educational institute. With moving focus on information and communication technology (ICT), all the renowned universities and institutes are launching online courses for students or executives who otherwise are not able to join them on a full-time basis.

- **Science and Technology:** Science and technology is another field in which computers are used extensively. Computers' fast and complex processing abilities are used in most research and development in the field of science and technology. Problems, which were previously highly complex in nature and were considered very difficult to resolve, have been rendered possible by using a computer. Architects and engineers use computers regularly in their work. Satellite communication has become a reality today only because of the role played by computers in it. With the help of computers users are able to communicate in remote areas where telephonic communication is not available. Simulation techniques are effectively provided by computers. Simulation is a process by which an artificial environment is created and is a replica of an actual environment in which a user needs to work or compute. For example, astronauts training on Earth can use a space-like environment.

- **Medicine:** Medical research and diagnosis is an area that is extremely influenced by the use of computers and the tools provided by it. Be it the development of new medicine, new techniques, or records of patients to be maintained in hospitals, the entire administrative work, which includes imparting information about the number of rooms' available, detailed information about the patients, etc., is computerized. Computers are also used in preparation of certain medical reports like ECG, CAT-SCAN, etc. Robotics is used for conducting complex surgeries on the patients.

- **Law and Order:** Another field where computers have found their beneficial usage is law and order. With the help of video conferencing, the court can hear the cases of criminals without moving them from jail to court. In some countries, CCTV cameras are installed at sensitive places to keep a close watch on the movement of people. These CCTVs are monitored and controlled with the help of computer systems and are helping in maintaining the law and order in many cities. Supreme Court, High Court, and Lower Court are all computerized. One can log into the court site and find the status or resulting judgment of a case. Lawyers use computers for recording their cases and retrieving information about a particular case. Computers have proved to be very useful in crime detection techniques like fingerprint identification.

- **Business:** Businesses are undergoing a revolution in the way in which they perform. Today, even a grocery-shop owner has a website through which the user can buy household items without going to the shop. Computers are used extensively in the designing and manufacturing of products with the help of CAD and CAM. All business functions such as production, sales, payroll, inventory management, dispatch, etc., are now computerized. Computers are used broadly for doing sales analysis, forecasting, generating reports, and most recently in customer relationship management. Various routine clerical tasks such as maintenance of files, ledgers, etc., have been taken over by computers. Computers are also used for booking online tickets for cinema, railways, airlines, for online banking transactions, paying online bills, etc.

Thus one can say that there is hardly any field of any business application that is not influenced by computers. The impact of computers is such that if anybody refuses to accept computers, they are soon bound to become outdated.

Test Your Knowledge

1. What do you understand about data, processing, and information? Explain with an example.

2. Explain the meaning of the term "processing."

3. Are the words data and information interchangeable? If yes, then explain with an example.

4. Why is there a need for data and information? Explain.

5. What meant by a "computer system"?

6. Can a computer system be called an information processing system? Explain your answer.

7. Discuss the various types of a computer system.

8. Discuss the various generations of a computer.

9. Discuss the various components of a digital computer system.

10. Discuss the basic architecture of CPU and its functions.

11. Discuss the block diagram of a computer system.

12. What do you understand about input devices? What are the various input devices, and which of them can be used for the purpose of entering data into a computer system?

13. What do you understand about output devices? What are the various output devices, and which are used for the purpose of showing output of a computer system?

14. What is meant by auxiliary storage or auxiliary memory? Discuss the various storage devices used in a computer system.

15. What do you understand about offline storage devices? What are the various devices used for the purpose of storage?

16. What do you understand about the backup storage devices? Explain in detail.

17. What do you understand about memory? How many types of memory are there? Explain.

18. What do you understand about the peripherals used in a personal computer?

19. What do you understand about the processing speed of a computer system? What are the factors that influence the performance of a PC?

20. Can a PC be described as a virtual office? Explain your answer with an example.

21. Explain the factors that govern the processing speed of a computer system.

22. What are the various usages of a computer system for the purpose of business?

CHAPTER 2

SOFTWARE: AN INTRODUCTION

2.1   Software: An Introduction

Software is the program that instructs a computer how to process data and generate required output.

2.2   Types of Software

Software can be divided into two categories:

System Software: A computer user does not understand the machine language (a language that a computer understands), and similarly a computer cannot understand a high-level language in which a user communicates. So, if a user wants to work on a computer system and wants to run an application program, the computer needs to have system software. *System software* can be defined as a collection of programs that performs the following functions:

- Receiving and interpreting user commands (i.e., converting users' instructions into machine language and vice versa).

- Running application programs and storing them in the hard disk or any other secondary storage as per the directions given by the user.

- Retrieval of the stored programs from the hard disk or any other secondary storage device on the user's command.

- Creates an interface among the peripheral devices and the CPU, directs and produces the results on the user's command.

Thus, it can be said that system software is responsible for the coordination of all activities in a computer system.

Application Software: *Application software* is written with a specific purpose in mind. Application software allows users to work in English or give the commands in a format that is not dependent on computer hardware. To run application software there has to be system software. Application software includes high-level language programs like basic, C, C++, Pascal, etc., or packages like Microsoft Office, Lotus Smart Suite, etc. It is not necessary for the high-level programmer to know the machine-level programming. Application software can be further classified into two categories.

- The software development firms like Microsoft Corporation, Oracle etc., prepare general-purpose software. Software included in this category are Tally, Ex from TCS, Microsoft Office from Microsoft Corporation, and many more.

- Tailor-made application software is prepared by software development firms according to the needs and wants of their clients. For example, software development prepared by Infosys, Wipro, TCS, and many more software development firms for their clients.

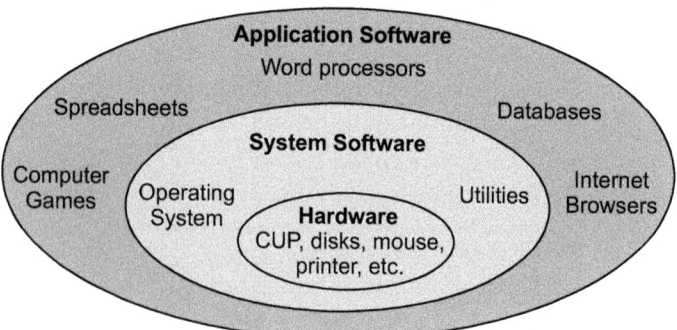

2.3   Assembler/Interpreter/Compiler

Programming languages can be classified in the three categories. They are as follows:

- *machine language*
- *assembly language* (LLL)
- *high-level languages* (HLL)

Machine Language

These are the languages that are machines dependent. Machine language is directly understood by the computer system, as it consists of a series of binary digits (0s and 1s); therefore no conversion is required and the processing of the program is very fast. Because the instructions given in a machine language are in binary digits only, it is very difficult for the common user to write a program or work on the computer system; it is also a very time-consuming process. Not only do the programmers need to remember the codes representing the machine instruction set, but they also have to keep track of the storage location of the data and the instructions. It is very difficult to detect and rectify errors in a machine language program. For this reason, these languages are not popular.

Assembly Language (Low-Level Language)

Because of difficulties previously explained, a machine language is not widely used, and thus assembly language comes into existence. These languages were developed in the early 1950s and used mnemonic operation codes. For addresses, symbolic representations were used. The first step in programming involved replacing the numeric binary machine-language operations codes with the mnemonic names. Machine codes were still in use for processing data; therefore special software was required, which translates the programs written in assembly language into a machine language. The major limitation of the assembly language was that they were machine dependent, that is, programs written on one machine will not work on another machine.

Assembler: *Assembler* is a special software that converts the program in assembly language into machine language. The program written in assembly language is called *source code* and is converted into a machine-readable format by an assembler known as *object code*. Assembly languages have many advantages over machine languages, as they save time and reduce complexity. The number of errors committed is also reduced and the identification and removal of bugs becomes quite easy.

High-Level Languages (HLL)

Working with assembly language was easier in comparison to a machine language, but still the common user had several difficulties in writing programs; thus the high-level languages (HLL) were developed. With the advent of HLL, working on a computer system became an easy job for the common user. These languages are very close to the English language, and they move away from machine dependency. Unlike the low-level languages, all instructions are written in English, which follows some rules called syntax of the language. But computers understand only machine language, therefore these high-level languages need to be translated into the machine code, and the translator used for this purpose is an interpreter or a *compiler*. In low-level languages, only one instruction was translated into a machine instruction at a time; however high-level language statements are translated into several machine code instructions by the translators.

Interpreter: An *interpreter* can be defined as software that is used for converting a program written in a high-level language into machine language. The interpreter takes the high-level language program code as input, line-by-line, and converts it into a machine language, line-by-line. No object code is produced in this process. If the interpreter finds any error in any line of the program, it stops there and notifies the user about the error. The user needs to correct the error and then rerun the interpreter. The translation of the program from the high-level language into machine code restarts from the beginning of the program, thus making the conversion of the program a time-consuming process.

To overcome this problem, another translator program called *Compiler* was designed.

Compiler: It can be defined as a translator program that converts the high-level language programs (source codes) into machine languages, and the object code is generated, stored, and can be executed at any time. *Object code* refers to a code that can be run to produce the output of the program written in a high-level language even without the availability of the actual program (source code). In the process of compilation, the compiler converts the complete program into an object code (machine readable) and notifies the user about all the errors in the program. Now the user needs to go to the source code (the actual program written in HLL) and rectify all the errors reported by the compiler, and then he needs to recompile the program. The recompilation process is a must, otherwise new corrected object code will not be generated and the user will have the earlier incorrect object code in the system, which will still produce errors. Compiler programs are very effective in comparison to an interpreter, as they take the whole source program and convert it into reusable object code which results time savings. Now, almost every high-level language comes with its own compiler, which makes them very efficient.

2.4   Generations of Computer Languages

Evolution or generations of programming languages are described as follows:

- **First-Generation Languages:** The first generation of languages started around 1940. The programs were written in machine languages that were very difficult to code. Later in the 1950s, programming in assembly languages started, and the assemblers were designed, which converted the assembly language code into the machine-level code.

- **Second-Generation Languages:** These languages came into existence between the late 1950s and early 1960s. These languages led the way for the introduction of many new concepts in programming. ALGOL60, COBOL, LISP, FORTRAN, etc., were some of the popular languages of the second generation. The concepts of data type and structured programming were also developed during this generation.

- **Third-Generation Languages:** Third-generation languages saw their evolution in the late 1960s and early 1970s. Some of the popular languages of this generation are SIMULA67, BASIC, SNOBOL 4, C, and PL/1. The concept of record structure and classes, arrays, pointers, storage classes, etc., were also introduced during this period.

- **Fourth-Generation Languages:** Most of these languages had their base in third-generation languages and evolved in late 1970s. Many software development tools were introduced which enhanced the productivity of fourth-generation languages. These languages interact with database management systems (DBMS) tools for storing, manipulating and retrieving data.

High-level languages are usually considered to be the procedural languages; on the other hand most of the fourth-generation languages are nonprocedural languages. *Procedural language* means a programming language that requires writing a series of systematic statements, functions, and fixing syntax to complete a program, for example, COBOL. Most of the statements in fourth-generation languages are simple and self-explanatory, which made them popular and widely used. However, fourth-generation languages also provide a user facility of just specifying the output without writing an actual program, wherein the program is automatically generated by the software. For example, the recording of a "MACRO" where the user simply records the steps and the program is generated by the software itself.

2.5   Uses of Computer Languages

Various applications of programming languages are as follows:

- **Scientific-Mathematical Applications:** These are programming applications that use the concept of science and a high level of mathematical computation. These programs are written to fulfil the mathematical, scientific, and statistical programming needs of the user for solving problems with the help of a computer. These programming applications are usually complex in nature, and to use these programs, users need to be well versed in mathematical principles, algorithms, and statistical principles so the problems can be solved properly by writing a program. A couple of examples of this are SPSS and R-Software for statistical analysis.

- **Text Processing Applications:** When a user needs to draft and format a letter, text processing applications are required. These applications involve manipulation of any natural language text as data. Word processors like MS WORD are included in this category. These text processing applications help to enhance the productivity of a business organization.

- **Data Processing Applications:** Today, data is considered an important asset for an organization. Organizations need to record, store, maintain, summarize, and retrieve data on a day-to-day basis. To fulfil this need, many data processing applications have been developed. The level of data involved in daily processing is very high and its management is very important. Therefore the data processing applications that have been developed can support huge levels of data transactions.

- **System Programming Applications**: System software is required to act as an interface between the user and the computer. This is because a computer system understands only the machine language that is not understood by the user. System software helps the user to work in any high-level language and converts it into the machine code that is understood by a computer. The operating system, compilers, interpreters, schedulers, etc. are included in this category of applications. "C" language is used to write system-programming applications.

- **Artificial Intelligence Applications:** Artificial intelligence means a situation where a computer can think like a human. The applications of this concept are varied and include logical games like chess, bridge etc. Most of these applications use LISP and PROLOG languages, which are logic based.

Advantages of High-Level Languages:

- easy to learn

- easy to understand

- easy to program

- easy to maintain

- easy to document

- easy to debug

- less time-consuming

- portable

Here, *debugging* means the removal of errors in the program, and portability means when a program is written in a high-level language, it can run on any computer irrespective of the hardware configuration.

2.6   Operating Systems

2.6.1 Definition

An *operating system* is system software. It is a set of programs that provides an interface between the user and the computer system (hardware). In other words, it coordinates the flow of information from the computer to the user and vice versa.

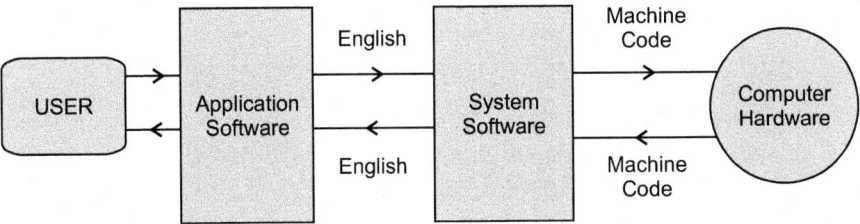

2.6.2 Functions of an Operating System

The operating system performs a number of services and/or the functions for the users of the computer. These operating system (OS) services or functions that might be performed are as follows:

- **Memory Management:** Here memory means random access memory (RAM). When the programs are loaded into the memory and are executed, the OS reads them from the hard disk and loads them into the memory (RAM). But before it loads, the OS checks whether the memory is available or not, and if available, it allocates it to the program. Once the program execution is over, the OS removes the program from the memory (RAM) and this freed memory can be used for another program.

- **CPU Management:** A microprocessor executes a number of processes at one time. Deciding which process is to be executed first, and which next, is a complex problem. This problem is effectively handled by the OS and is known as *job scheduling*. The OS helps the processor to schedule its activities so as to fulfil the requirements of various parts of the computer system. The processor gets the job from the memory, processes the job, and passes the result onto the predetermined place and makes itself ready for the next job.

- **Disk Management:** Data and program storage on the hard disk or any other storage device is a complex phenomenon. It is very difficult for a common user to understand this. Thus, the OS spares the user from understanding this process as it coordinates the storing and retrieval of files.

- **Input/Output Management:** Numbers of peripheral devices such as keyboard, mouse, printer, projectors, etc., are attached to a computer system. To make these devices work properly, the system requires supporting software called *device drivers* in the memory. The device drivers are also known as the *support utility programs* of the OS and are controlled by it. All these devices communicate with the computer for obtaining the job and getting the job done, which is again monitored by the OS.

- **User Interface:** The system, but not the user, understands machine language, which consists of 0s and 1s. The OS bridges the gap between the machine understandable language and human understandable language by providing a command interpreter.

- **Communication:** A computer system runs a number of processes at a given point of time and there exist situations where one process needs to exchange information with another process. The OS facilitates these communication processes within a system and among the systems. Communication is done either via shared memory or by the technique of *message passing* in which packets of information are moved between processes by the OS.

- **Error Detection:** While working on a system, it is quite possible that a user may commit a number of errors. The OS constantly keeps track of all possible errors. Errors may be typing a wrong file name, wrong syntax of a command, network failure, a printer jam, or running out of paper. These are errors that are frequently committed by a user. For each type of error, the OS takes an appropriate action by showing an appropriate message on the computer screen prompting the user to correct their action.

- **Resource Allocation:** The OS does the proper allocation of computer resources as there might be multiple users or multiple jobs running at the same time. Resources like processing time, main memory (RAM), file storage, devices, etc. must be allocated to all of the users equally, and this is handled by the OS.

- **Accounting:** The OS keeps full track of how much and what kind of computer resources are used by a user. This record keeping can be used for improving computing services.

- **Protection:** The OS provides a multilevel protection mechanism to users. This protection is required by users from other users or an outsider. The OS provides a mechanism by which a user can protect his data, and this is done by controlling the access to system resources. For this, the OS uses the authentication process in which each user has to authenticate himself to the system. This is done by means of a multilevel password protection system provided by an OS.

2.6.3 Types and Classifications of Operating Systems

System software can be broadly classified into the following categories on the basis of their usage:

- **Batch Processing System Software:** A negligible interaction between the user and the program *batch processing system software*. In this type of system jobs are processed in the order in which they are entered, that is, on a "first in, first out" basis (FIFO). In a batch processing system, memory is divided into two parts; one is permanently occupied by the software, whereas the other is used as per the need of the user. It simplifies the processing operations because the instructions are executed in batches, and thus saves the processor time.

- **Multi-User Operating System Software:** The *multi-user operating system* supports the multiple units of PCs called "terminals" that are attached to the main computer system as in mini and mainframe computer systems. It consists of only one central processing unit (a microprocessor) that performs all the operations. These systems are used when two or more users try to run programs at the same time. Examples of the multi-user operating system are UNIX, MSV, etc.

- **Multiprogramming or Multi-Tasking System Software:** This is the system software that is capable of running more than one program at the same time. *Multiprogramming* can be defined as a process of creating a situation in which more than one program may be held in the main memory at one time, thus making it possible to process several programs at a time. The main objective of developing this kind of system software is to minimize unused microprocessor time. A computer switches from one job to another at a rapid rate under the time-sharing mode. Different terminals are used to enter jobs into the computer. After processing the first user's job, it proceeds to the second and then to the third and so on for a short period of time called the "time slices," before returning to the first user's job from where it earlier started. This cycle continues indefinitely. When one program is finished the other program replaces it. UNIX, OS/2, and Windows are commonly used multiprogramming or *multitasking operating systems*. The processor is kept busy while channels and buffers are occupied with a job of bringing data and writing out information.

For example, let us assume that three users are working on a system simultaneously. In this concept the program of each user will be divided into a number of pages (layers) of equal size. During execution, the processor will divide its processing time equally among all of the users. It will first process the layer 1(L1) of the *program 1* (P1), then *layer 1* (L1) of the program 2 (P2), and finally the layer 1 (L1) of the program 3 (P3). The processor will give equal time to all the users, but it will appear to all of the users as if processor is giving its full time to them. When program 1 (P1) is finished, the processor will divide its time equally among the remaining programs (P2 and P3) and finally to the program 3 (L3).

P1, L1
P2, L1
P3, L1
P2, L2
P3, L2
P3, L3

- **Multiprocessing Operating System Software:** In a *multiprogramming operating system,* more than one program is processed by an operating system, whereas in a *multiprocessing operating system,* one program is processed by more than one processor. *A multiprocessing operating system* software uses multiple processors that share a common memory. Instructions from different and independent programs are processed at the same time by different processors. On the contrary, the processors may simultaneously execute different instructions from the same program. Examples of commonly used multiprocessing operating systems are OS/2, UNIX, MSV, etc. Multiprocessing systems can be classified as:

  - loosely coupled multiprocessing

  - functionally specialized processors

  - tightly coupled multiprocessing

  - parallel processing

Multiprocessor systems usually consist of two or more processors. Each processor has its own CU, ALU, etc. An interconnection mechanism allows each processor to access shared main memory and input/output devices. The processors not only communicate with each other through memory but also are able to directly exchange signals. Memory is organized in such a manner such that it provides a multiple simultaneous access to a separate block of memory. The operating system controls this entire system and provides interaction between processors and their programs.

- **Real-Time System Software:** Real-time systems are the systems in which response time is critical. These are the systems that are involved with the immediate processing of data, machines, and records. These systems are designed to accept the data in real time, which means as soon as an activity occurs, the system processes the data immediately and generates the output in time to have an effect on the ongoing activity. Real-time systems are online systems with tighter constraints on response time. Examples of the real-time operating systems are C Executive, communications control program (CCP), CTOS, CTRON, FADOS, etc.

2.6.4 Components of an Operating System

An operating system consists of two primary components:

- a supervisor

- an integrated set of support utilities

1. **The Supervisor (or Control Program):** The *supervisor* is defined as the component of an operating system that takes care of the overall working of the computer system. It is a set of programs that are integrated to one another. It performs the following three basic functions:

   - It initializes the system at the time of start-up

   - It allows running of the application programs

   - It controls input and output devices attached to the system.

The supervisor also performs some additional functions such as keeping a track of computer time for different users, etc. The OS is generally found on a hard disk and sometimes in the form of a chip called *firmware*. The supervisor consists of two portions: a *kernel* and a *transient* portion. When the supervisor is loaded into the memory for the first time, both portions are loaded. The kernel part of supervisor always remains in the memory with an application program. This is to monitor the system operations. It's the part of an operating system that directly interacts with the hardware, and therefore it must be present in the memory as long as the computer is being used. The transient portion need not always be present in the memory. When executed, a program may overwrite this part of the OS. Once the program execution is over, the transient portion is reloaded into the memory.

2. **The Support Utilities:** The *support utilities* are the system programs that perform useful functions. They are classified as follows:

   • *Program Development Utilities*: These include assemblers, editors, interpreters, compilers, and linkers.

   • *System Management Utilities*: These include the programs that keep track of more than one user, diagnostic routines, etc.

   • *File Management Utilities*: These include programs for copying files, erasing files, printing files, renaming files, etc.

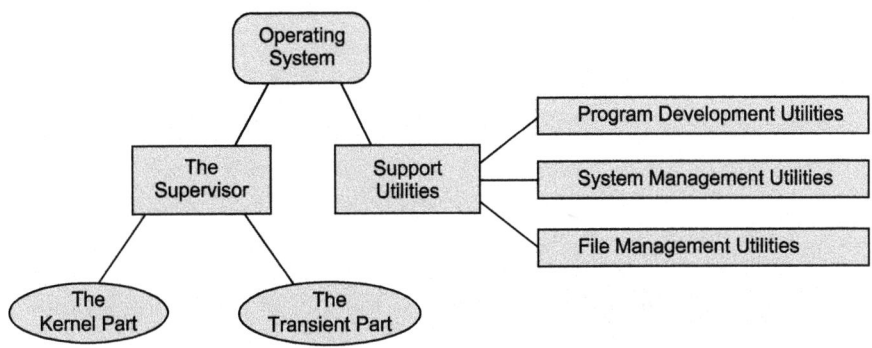

2.7   What is Graphical User Interface (GUI)?

Microsoft Corporation brought a revolution in the area of system software development when they marketed software called Windows. Prior to the launch of Windows it was not possible for the user to work on a computer with the help of graphics. What the disk operating system (DOS) gave users was termed command line interface (CLI), where users were required to type the right command with the help of the keyboard on the command prompt (C:>_). Remembering the exact syntax and spelling of the command was a tedious task; however with Windows the concept of a *graphical user interface* (GUI) was introduced. With a GUI, the user did not get the black screen of DOS with the pointer movement limitations; rather, users started to get a graphical screen where they were not required to remember the typical DOS commands for the purpose of doing their work. Here they were supposed to simply click on a picture (icon) option with the help of a mouse (a pointing device) and the Windows software automatically executed the command selected by the user.

The introduction of Windows brought a revolution in the field of computers and more and more users started to work on computer systems because it became easier to operate a computer system with the help of the Windows software. Before Windows 95, the earlier versions launched for Windows were Windows 3.1 and Windows 3.11 (network), but these required DOS to run. Windows 95 is in itself an operating system with a facility of GUI. Various versions of Windows OS launched so far are Windows 98, Windows 2000, Windows ME, Windows XP, Windows 7.0, Windows 8.0, Windows 8.1, and latest in the series is Windows 10.

2.7.1 Elements of GUI-Based Operating Systems

The Windows OS has various advantages over the DOS operating system and these advantages are:

- It is very easy to learn.

- It allows the user to work on multiple applications simultaneously. For example, one can type a letter in the foreground and in the background, printing can take place.

- It supports networking with other computer systems.

- It contains many advanced built-in features that were not available in DOS.

- Windows is user-friendly with a facility of GUI, that is, commands are represented by pictures and there is no need to remember the right syntax of any command.

- Windows offers common menus, that is, the same command in many related softwares does the same job (for example commands to save and print).

- Windows provides a facility to transfer data between different applications, whereas this is not possible in DOS.

Directory Management in Windows

■ **File**

A *computer file* is an entity that contains text, record, or a specific piece of data. A computer file may be anything from an executable program to a user-created document. A computer file consists of a file name that has two parts: one is a primary file name that creates the file's identity, the other is an extension or a secondary file name that tells the operating system and associated programs what type of file it is. In DOS, the name of a file was limited to eight characters, but modern Windows systems allow for much longer file names. The primary file name is generally determined by the user and defines the nature of files. A secondary file name or the file's extension is typically a period followed by two to four characters. This part of the file name is used by the operating system for internal cataloging. The way an extension is used varies between operating systems; some require the extension and some completely ignore it. The extension helps identify the programs used with the file and acts as a shortcut when read. For example, ".ppt" means it is a PowerPoint file. Some of the important characteristics of a file are:

- All files stored in a system have a size, although it might be zero.

- A file is stored on a location in a computer system. This is maintained by creating an index by the system for all files stored.

- Certain attributes are associated with every file like read, write, archive hidden, etc. These attributes can be set by right-clicking on the file and clicking on the "properties" option.

- It is very easy to create, delete, or modify a computer file.

Some of the Computer File Types by Extension

Extension	Associated Program	Extension	Associated Program
AVI	Animated Video	MID	MIDI
BAT	DOS Batch File	MP3	Audio
BMP	Bitmap Graphic	PDF	Adobe Acrobat Document
COM	Program	PIC	Bitmap Graphic

DAT	Data file	ZIP	Compressed Archive of Files
DLL	Dynamic Link Library	PM4	PageMaker 4
DOC	Microsoft Word Document	PM5	PageMaker 5
EXE	Program	PPS	Microsoft PowerPoint
HTM	HTML Hyper Text Markup Language	PPT	Microsoft PowerPoint
ICO	Icon	RTF	Rich Text Format
JPG	Image	SYS	System Files
TMP	Temporary File	WAV	Audio File
TXT	ASCII Text	XLS	Microsoft Excel Spreadsheet

● **Directory**

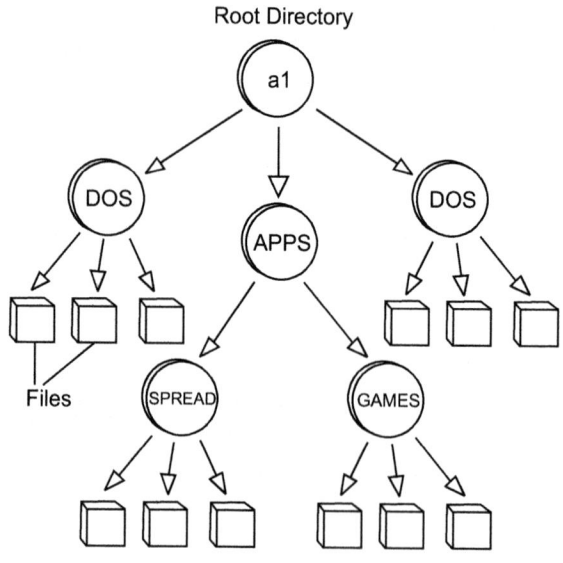

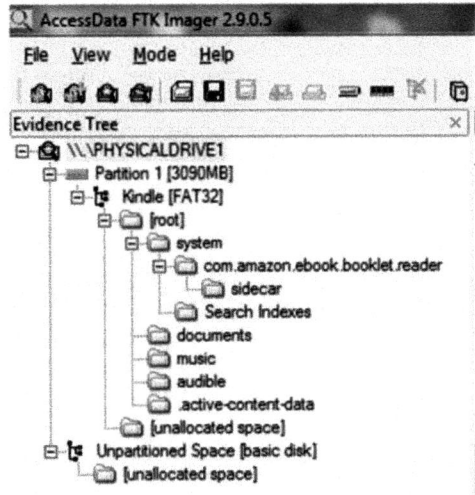

A directory commonly called *folder* can be defined as a place or a container that can be used to store files or sub folders. A user can consider a *directory* as a file cabinet that contains folders in which files can be kept. A directory is often defined in terms of an *inverted tree*. In Windows it is very easy to access a file in a directory. The topmost directory is called the *root directory*. A directory that is under another directory is called a *subdirectory*. A subdirectory may contain files or another subdirectory. A directory above a subdirectory is called the *parent directory*. In Windows the term *directory* is replaced by folder and subdirectory is by subfolder, Under DOS and Windows, the root directory is a back slash (\).

■ **Creating Directory/Folder**

A directory (folder) is used to store files. A user can create a number of folders and subfolders inside a folder as per the need. Steps to create folder are as follows:

1. Right click on the blank area of desktop and right click the mouse.

2. A drop-down menu will appear. Click on the "New" option and under New on the "Folder" option as shown in the following figure. This process will create a folder on the desktop labeled "New Folder," however the name can be changed by the user.

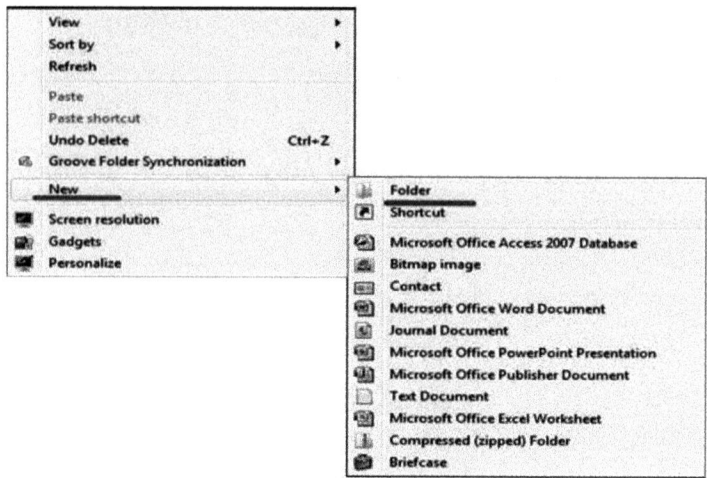

3. A new name can be typed for the new folder, for example in our case it is "Word Files" and then press the Enter key.

- **Creating Subdirectory/Subfolder**

A subdirectory/subfolder is a directory/folder that is created in an already existing directory/folder. Users need to follow the same steps as taken earlier to create a folder with only one difference. While creating a folder, the user has to click on either the desktop or on the desired location instead of creating a subfolder. The user first needs to move inside the folder by double clicking on it and then repeating the steps previously discussed to create a subfolder.

■ **Renaming a Directory or Folder/File**

The user needs to take the following steps to rename a directory/folder:

1. Right click on the folder; a drop-down menu will appear. Click on the rename option as shown in the following figure.

2. Change the name of the folder as per your choice.

 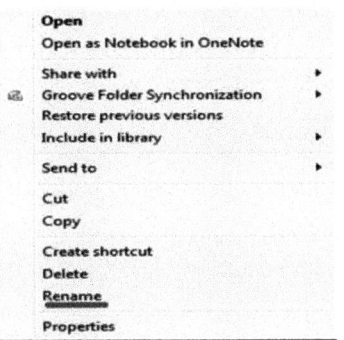

3. The folder will appear with a changed name as shown in the following figure:

■ **Copying the Directory or Folder/File**

Copying a directory/folder means creating a duplicate of the original. The following are the steps for copying a directory/folder:

1. Open the location that contains the folder you want to copy. In our example, the folder exists in the disk partition E with a name "AAGAAZ RBSMTC."

2. Right-click on the folder and click on the "Copy" option from the drop-down menu.

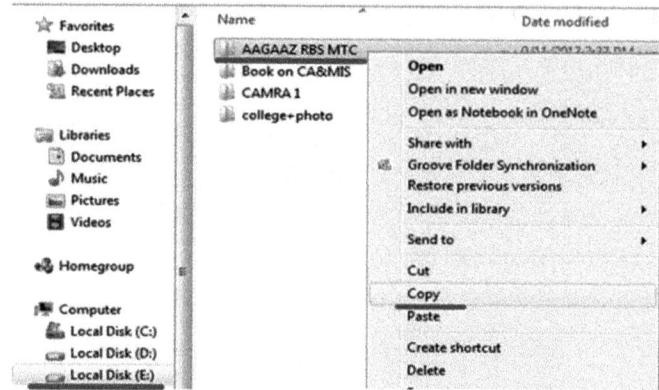

3. Open the location where you want to store the copy. In our example, it is disk partition D. Now right click anywhere and then click on "Paste" option as shown in the following figure:

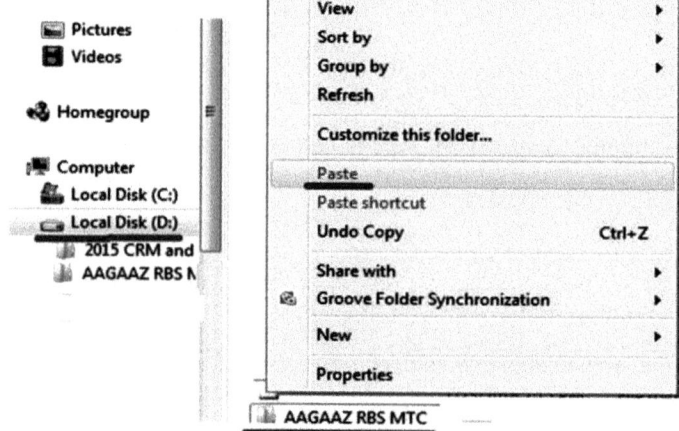

The copy of the original file or folder appears at the new location.

- **Deleting the Directory or Folder/File**
  - *Delete key*: The easiest method of deleting files in Microsoft Windows is by left clicking on the file, folder, or shortcut the user

wants to delete. The folder, file, or shortcut will now be highlighted. Pressing the delete key on the keyboard will delete the particular file, folder, or shortcut.

- **Delete file by right clicking:** Right click on the file, folder, or shortcut you want to delete, a drop-down menu will appear as shown in the below figure. Click on the delete option to complete the action.

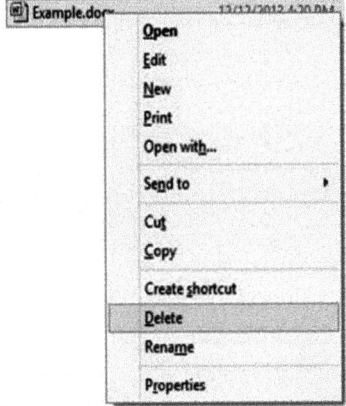

■ **Creating a File**

A file may consist of text, data, or records and is to be stored by assigning it a file name. A file name can be created by following the steps shown here:

1. Click the "File" menu in the program you are using (for example Microsoft Word, Excel, or PowerPoint), and then click "New." This opens a new document.

2. When done working in the file, click on the File menu again, and then click Save As to name the file and save it to the desired location on the computer.

Another way to create a file on your computer is:

1. Navigate to the folder or desktop, you would like to create your file, e.g., "My Documents."

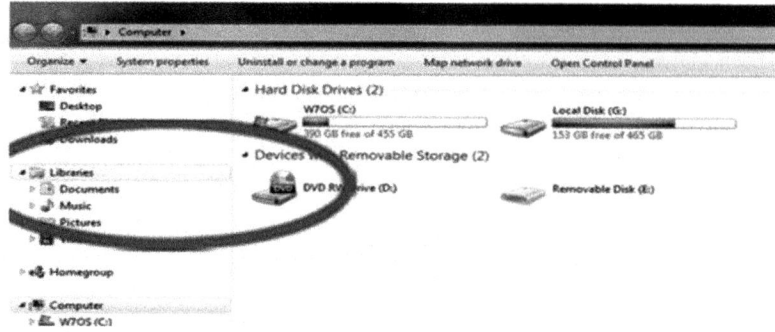

2. Right click on empty section of the desktop:

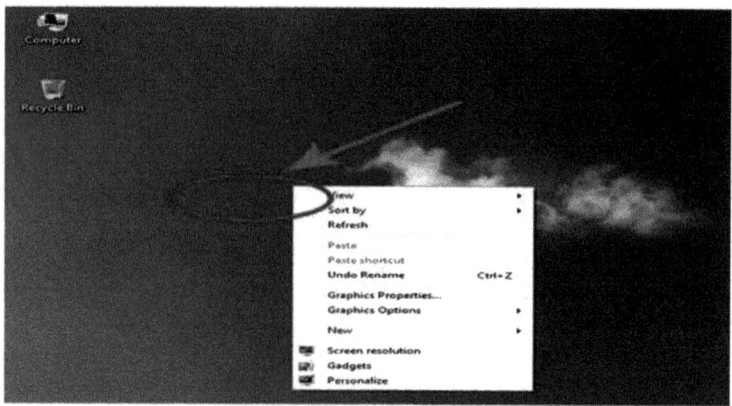

3. Select "New" from the menu:

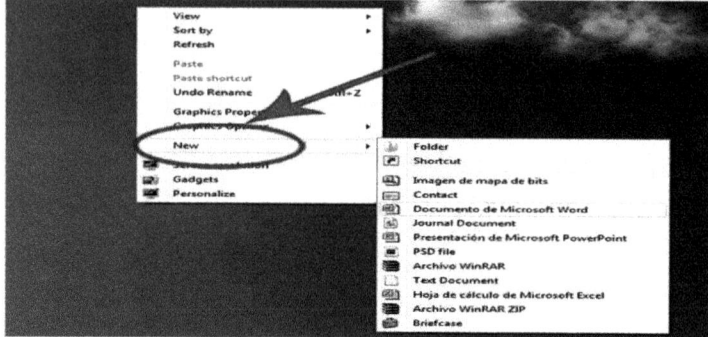

4. Select the type of file you would like to create:

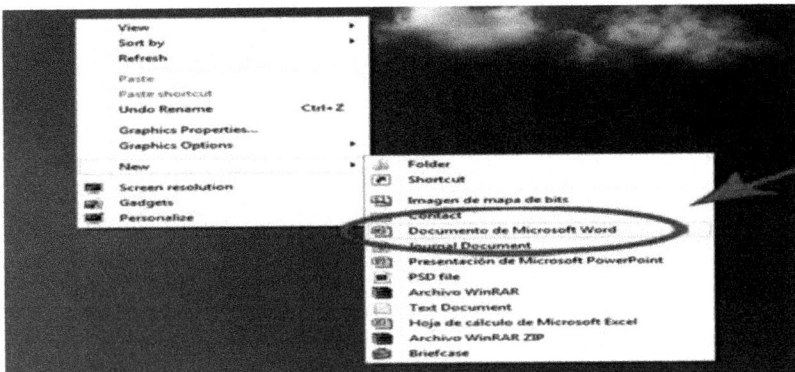

5. Entre the name of the newly created file:

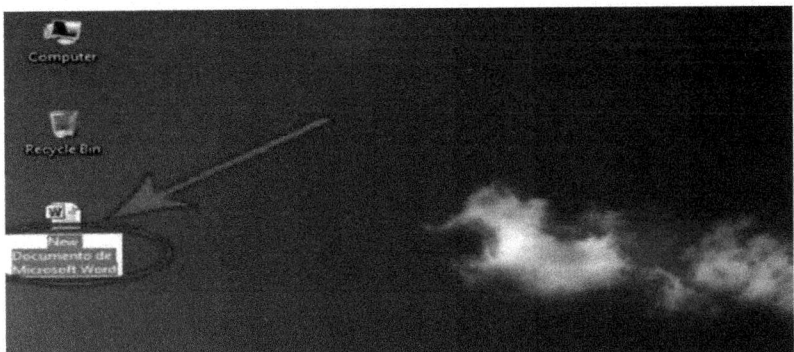

6. Open the new file to start work:

■ **Use of Menus, Tools and Commands of Windows**

Windows Explorer

Widows Explorer is a utility provided by the Windows operating system that performs various functions for the management of the disk and file system. Some of the activities performed by Windows Explorer are as follows:

To cancel the last action within a program or in My Computer or Windows Explorer	To change the appearance of items in a folder	To rename a file or folder
To change which program starts when you open a file	To control access to a folder or printer	To copy a file or folder
To create a file type	To create a folder	To delete a file or folder
To determine how much space is available on a disk	To make a copy of a disk	To modify a file type
To move a file or folder	To name a disk	To open a document from within a program
To preview a file	To put a shortcut on the desktop	To quickly send files and folders to another place
To remove a program from the Start or Programs menu	To select multiple files and folders	To show all files and file name extensions

Apart from this, there are many more options that can be performed with the help of Windows Explorer.

Control Panel: *The control panel* is that utility provided by the Windows operating system, which helps in setting the working environment for the user in which he or she wants to work. Various options provided by the control panel are shown in the following figure:

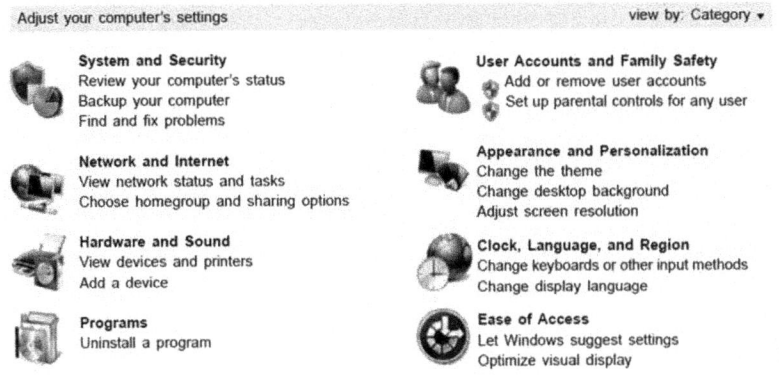

Options provided by the control panel allow the user to add new hardware, add or remove installed programs, set date and time, change the VDU display, set up fonts, set up game controllers, set up the keyboard, set up modems, set up a mouse, manage network/Internet settings, create a user's account, set up passwords, install printers, and many more options.

Print Manager: Print manager is the utility of Windows that helps the user in manage printer settings. To do this one has to follow these steps:

- Click on the "Start" option from the Windows opening screen.

- Click on the "Devices and Printers" option from the popup menu.

- Select the "Add a printer" option.

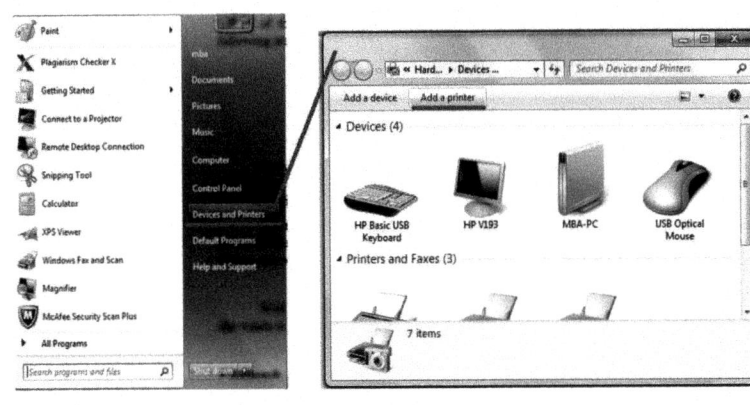

Paint: The "Paint" option provided by Windows is used for the purpose of creating, editing, and viewing pictures. The user can paste a Paint picture into another document such as Microsoft Word, or use it as a desktop background. Paint can also be used to view and edit scanned photos. To start working with Paint, the user has to take the following steps:

- Click on Start, point first to Programs, then to Accessories, and then click on the *Paint* option.

The Paint program provides various tools for making a painting on the drawing screen provided. Various tools provided for this purpose consist of color erase, fill with color, pick color, magnifier, pencil, brush, air brush, text, line, curve, rectangle, polygon, ellipse, and rounded rectangle. All of these tools are used for the purpose of creating drawings. Apart from these tools, the option also provides various color blocks below the drawing area. From these color boxes the desired color can be selected by the user for the purpose of using it in their drawing.

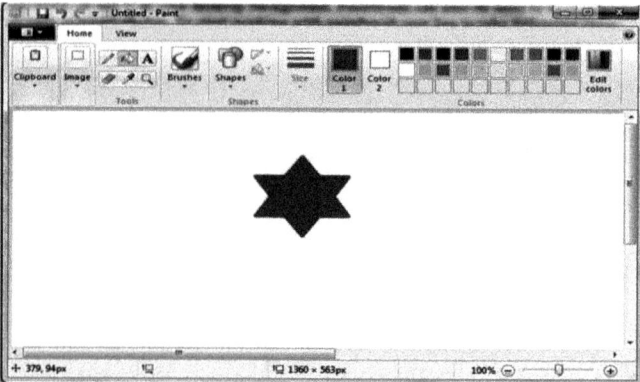

Calculator: The calculator provided by Windows is used for the purpose of normal mathematical calculation. To work with the calculator, the user has to take following steps:

- Click on Start, go to the Calculator option and click on it.

- Alternatively, if it is not visible there, then click on Start, go to the Accessories option, and under it click on the Calculator option.

Once all these steps are taken, a calculator appears on the screen and with the help of mouse or the keyboard the user can perform all the mathematical operations on the calculator. Calculator provides options such as normal, scientific, programmer, or statistics. The user can select the option and the calculator type will change accordingly.

| Normal | Scientific | Programmer | Statistics |

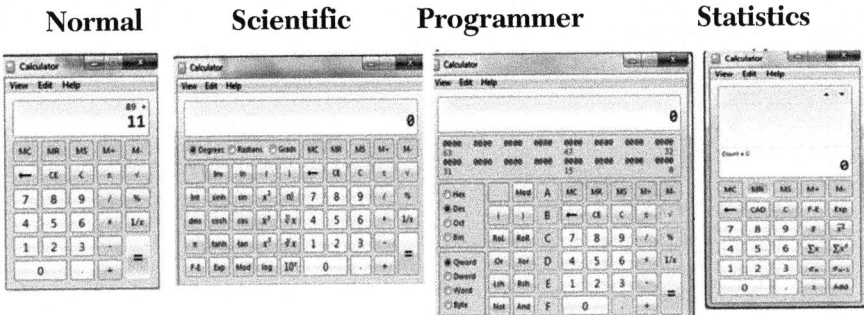

Desktop in Windows: The Desktop is the screen that appears once Windows is completely loaded. It is the screen from which users can start selecting options for their work. Active desktop makes it possible for the user to customize the desktop, launch programs, and switch between files. The desktop can be changed with the help of following steps:

- Right click the mouse button on a desktop; a drop-down menu will appear on the screen. From that menu select the "personalize" option to customize the desktop.

Taking following steps can make various changes to a desktop:

- The user can install desktop themes by clicking on "Control Panel," and then double-clicking on "Appearance and Personalization" option. Under this option there are numerous choices to change the settings of the desktop. Some of them are as follows:

  - change the theme

  - change desktop background

  - adjust screen resolution

  - add/uninstall gadgets

  - customize the start menu

Test Your Knowledge

1. Define software. Describe the many types of software.

2. Discuss the various functions performed by an operating system.

3. Explain the various steps involved in the booting process of a computer system.

4. What do you understand about programming languages? Explain the use of assembler, compiler, and interpreter with these programming languages.

5. Explain the development of the programming languages used by a computer system. For what purpose are assemblers, compilers, and interpreters used?

6. Define generations of programming languages. Discuss.

7. What is a machine language? How it is different from a high-level language?

8. Write down the various applications of programming languages with their advantages.

9. Explain multiprogramming and multiprocessing systems.

10. Explain the different types of system software available.

11. What do you understand about the classification of an operating system? Explain.

12. Explain the various components of an operating system.

13. Describe the meaning of a "file." How it is different from a "folder"?

14. What is a file name? Explain with an example.

15. What is a file extension? What is the significance of it?

16. What is a graphical user interface (GUI)? How it is different from DOS?

17. What is a directory/folder? Write down steps to create a directory/folder.

18. Explain how Windows Explorer in Windows operates.

19. Explain how the control panel in Windows operates.

20. Explain how print manager in Windows operates.

21. Explain how "Paint" in Windows operates.

22. For what purposes does Windows provide the calculator option?

23. Describe desktop in Windows.

CHAPTER 3

MICROSOFT OFFICE

3.1 Microsoft Office

Microsoft Office is an example of an application software that comes under the category of "packages" and is developed by Microsoft Corporation.

3.2 Working with MS Word

Microsoft Word is a word processor developed by Microsoft Corporation and was released on October 25, 1983. Microsoft Word is used for writing and editing text documents. It provides tools that help many users share and edit documents. Microsoft Word also provides its users with elementary desktop publishing capabilities and is the most popular and widely used word processing program in the market. Files created in MS Word are used for sending text documents via e-mail. Word 2007 introduced a new feature called *Ribbon* and later versions of WORD do more customization of it.

3.2.1 What is a Ribbon?

A ribbon in Microsoft Office is a feature that has replaced the file menu of the older versions. The following image shows how the ribbon looks. Images in the ribbon change on the basis of the "active tab," which means a tab that is presently selected by the user. For example, the ribbon for "home" is being displayed in the following image, and is the first option to appear when a user starts working. The ribbon only shows the options that a user wants to work with at any given time.

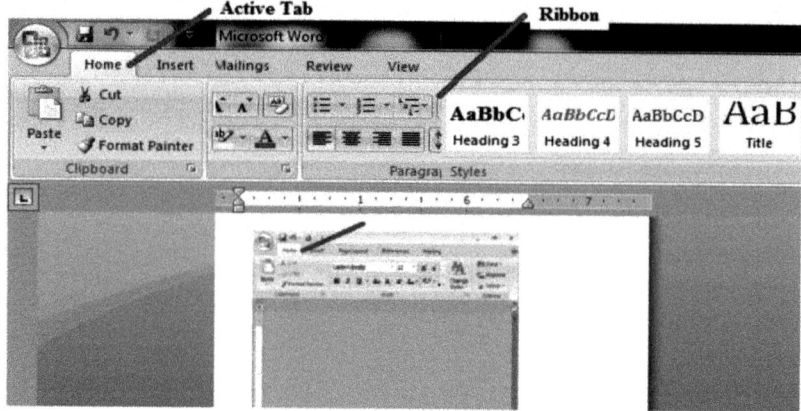

As the user changes the option, the ribbon appearance and options also change. For example, the following image shows the ribbon options of the "Page Layout" tab:

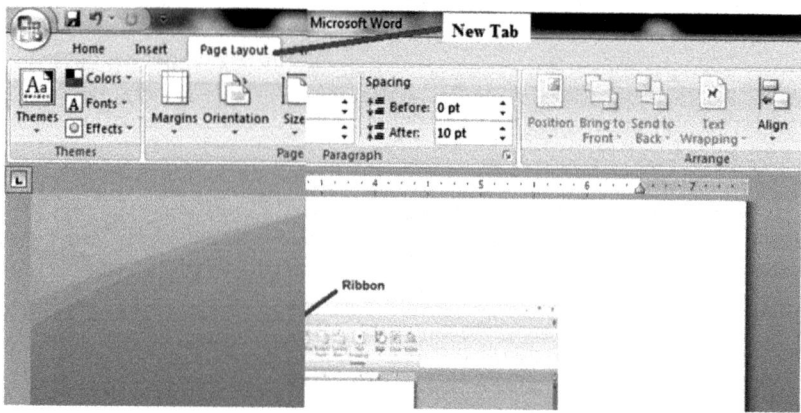

3.2.2 Office Button

Various tabs and options under them are as follows:

The pictorial option at the extreme left corner of the screen (shown in a circle) is known as an office button. This is actually the replacement of the file menu of the earlier versions of Microsoft Office. This button is found in

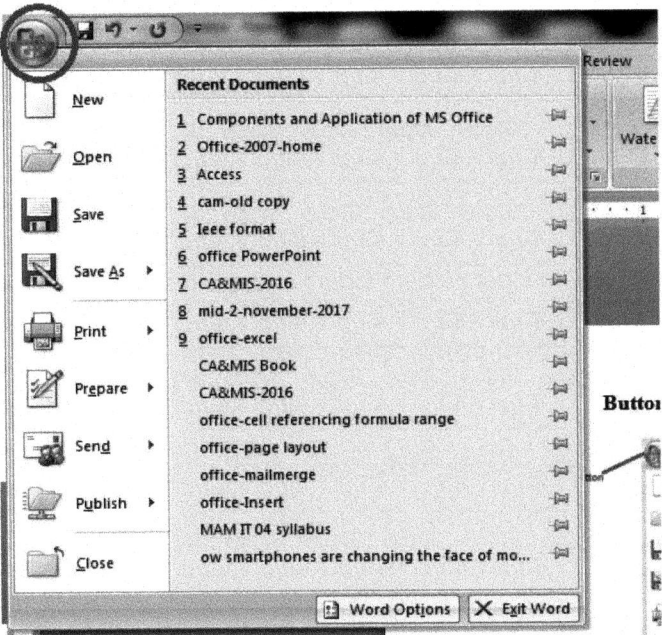

the programs Word, Excel, PowerPoint, Access, and Outlook. The working of various options under the office button are as follows:

New: This is the command that is used to open a new document. To open a "New" document click on the "office button" then click on "New." Lastly click on "Blank document" and then click on the *"create"* button.

Open: The Open command is used to open a file that already exists in the system. We can also open a file with the help of the shortcut key. The existing file can also be opened by pressing the shortcut key combination of "Ctrl+O".

Save: This command is used to save a document. For this, the user needs to click on the office button, then click the "save" command. A window will open where the user can determine the location where they want to save their file, the name of the file, and the file type. The file can also be saved by pressing the shortcut key combination of "Ctrl+S".

Save As: This command is used when the user wants to save the file with a different name, different format, and on a different location. This can also be done by pressing the shortcut key "F12."

Print: This command is used to get a printed page of the active document. The command gives three options to the user: (a) print, (b) quick print, and (c), print preview. Option "a" gives the user a chance to change the printer setting as per their needs. Option "b" sends the document directly to the default printer to save time, and option "c" gives the user a chance to see how the document will look after the printing and provides them a chance to still change the page settings.

Option A: Opens a print dialogue box

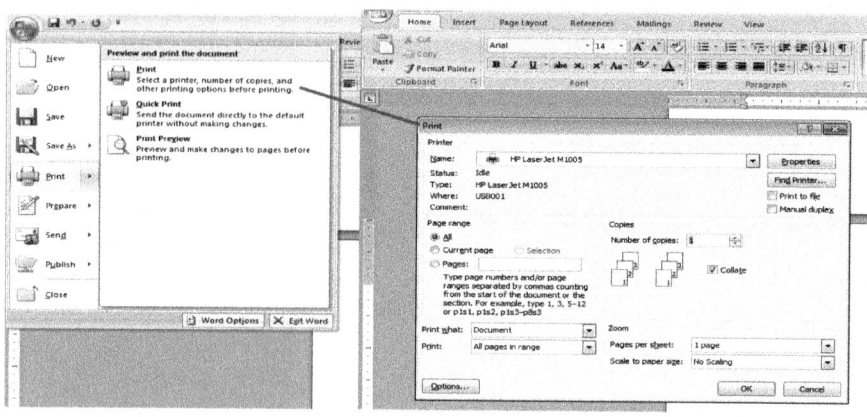

Ctrl+P directly opens the Print dialogue box on the screen

Option B: Offers the Quick Print option which sends the print directly to the printer.

Option C: Print Preview is an option which shows on the screen the actual printed page.

Prepare: In the *prepare* option a user finds eight sub-options. They are properties, inspect document, encrypt document, restrict permission, add a digital signature, mark as final, and run compatibility checker. In

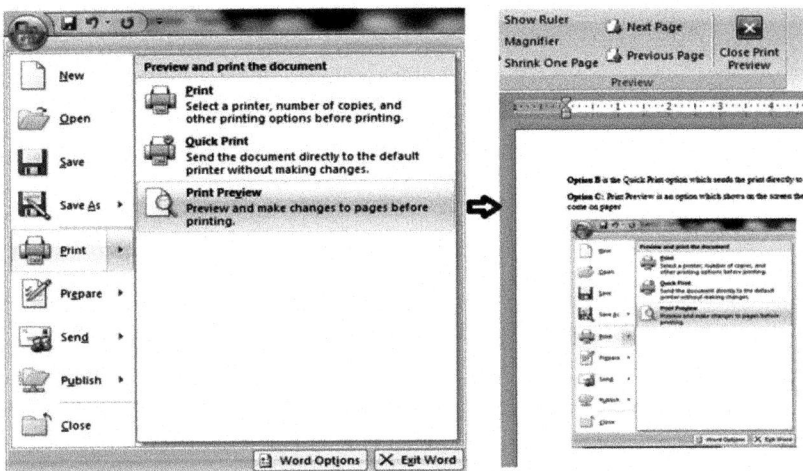

Word 2010/2013, this option has been changed to "Info" and under it only properties, inspect document, check accessibility, and check compatibility sub-options are available. The remaining four sub-options are found in the protect document option.

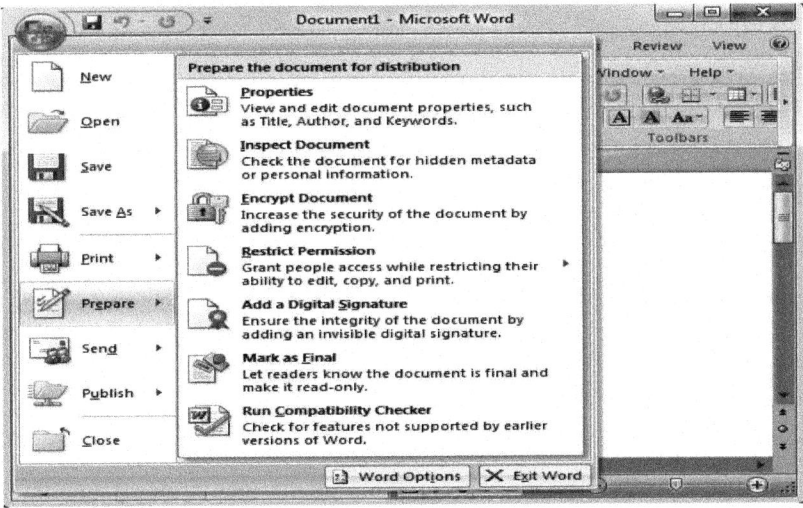

The *properties* option allows the user to edit document properties such as title, author, key words, and also if the user wants to add some comments to the document. In the *inspect document option* the user can check the document for hidden data and personal information if it is there. *Encrypt document* protects the document with a password. *Restrict permission* allows the author of the document to restrict the accessibility of the document from others. *Add a digital signature* will make the document authentic and authorized. *Mark as final* marks the document as a final copy and the user will not be able to make any further changes *Run compatibility checker* checks the compatibility of the document with other versions of the software.

Send: This consists of two sub-options: one is e-mail, which sends the copy of the document as an attachment to the mail. The second option is Internet fax, which enables the user to use Internet fax services to fax the document. However, for this the user needs to first sign up with the Internet fax service provider.

Publish: This helps the user to create a new blog post with the content of the document and to share the document by saving it to a document management server, and create document workspace by creating a new site for the document and keep the local copy synchronized.

Close: It closes the open active document and then Word. But if the user has made any changes in the document, it will first ask to save those changes and then will close the file, and then Microsoft Word.

3.2.3 HOME TAB

The "home" tab gives the user the following options:

Formatting Options

 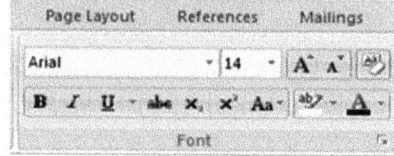

Cut: This command is used when a user physically wants to move all or some part of the text or image from one place to another in the document. To use this option, the user needs to first select the text that they want to move. This option also helps in the placement of text, image, tables, etc., from Word to Excel, PowerPoint, and vice versa.

Copy: The "copy" command works similar to the "cut" command, with the only difference being that the original position of text or image does not change, and only a copy of it is produced at the new location.

Paste: After using the "cut" or "copy" command, the material is placed on the clipboard. *Clipboard* is a temporary memory location to hold the material, including text, an image, or any other kind of a file. For the physical placement of the material, a user needs to place the mouse pointer at the desired location and click on the "paste" option.

Changing the size and font of the text: Font is a graphical representation of text that may include a different design that gives the different looks to the characters typed (called typeface), point size (the vertical length of the character typed, where there are approximately 72 points in an inch), the weight of the character typed, color, or design. Microsoft Word, Microsoft Excel, and Microsoft PowerPoint allow users to change the font, its size, and color when typing text in the document or spreadsheet or in a slide. Various steps to be taken are:

- **Step 1:** First select the portion of the text where you want to make the change with the help of the mouse. The selected text will become highlighted.

- **Step 2:** Click on the down arrow sign of the font style option and select the required font from the drop-down menu.

- **Step 3:** Now click on the down arrow sign on the box that contains numbers adjacent to the font style box and select the desired size of the font from the drop-down list of all available sizes.

Step: 2   **Step: 3**

The font size of the text can also be changed with the help of the buttons on the ribbon. For this the user first needs to select the text and then click on the up arrow sign to increase, or click on the down arrow sign to decrease the font button as shown in the following figure:

The B, I, and U Options

First select the text and then click on the **B** option on the ribbon. The text will convert to boldface. Select the text and click on the *I* option; the selected text will become italicized. Select the text and click on the U option, the selected text will become underlined.

The Strikeout, Subscript, and Superscript Options

First select the text and then click on the required option to get the desired result. The strikeout option is used to omit text. Subscript, also called subfix, refers to any character, number, or symbol written next to and below the text line of another character, number or symbol. The subscript usually appears in a smaller font size. On the other hand, superscript refers to a word, letter, number, or symbol written or printed above the text line of a word, letter, number, or symbol, and usually in a smaller size as shown in the example below.

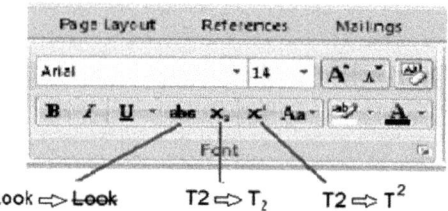

Changing the way a sentence looks: The following figure shows that on selecting the various options, the display of the sentence will change. But before selecting any of the options the user needs to select the text first, and then the display style of whatever needs to be changed:

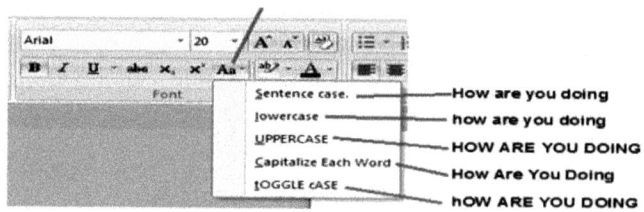

Changing the background color and the text color: First, select the text and then select option 1 to change the background color of the text. Execute option 2 for changing the color of the actual text.

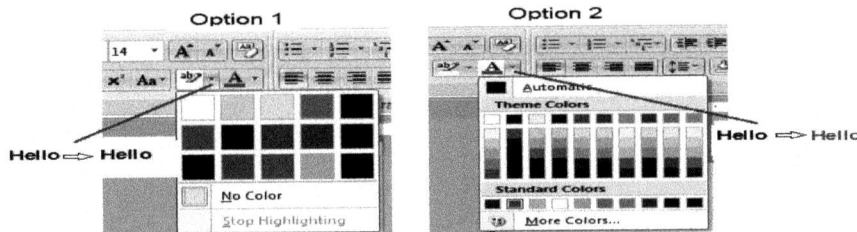

All these options can also be executed by clicking on the font option in the ribbon. A drop-down menu will appear from which the user can execute the desired command.

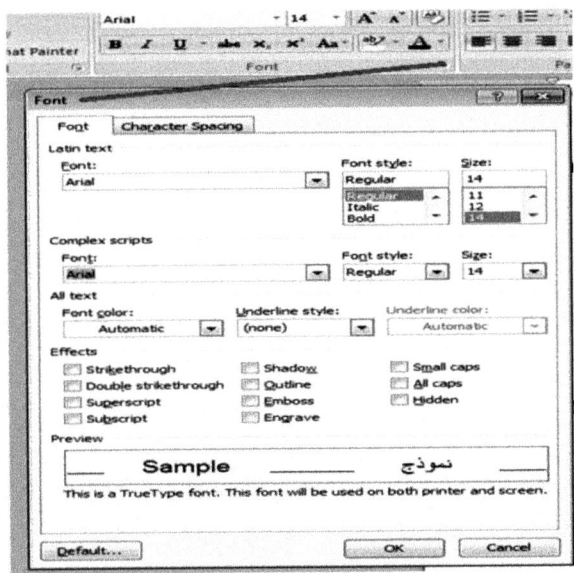

Formatting the Paragraph: A paragraph in Microsoft Word is any text that ends with an enter key. Paragraph formatting helps the user determine the appearance of paragraphs. For example, a user can change the alignment of text from left to center, or change the spacing between lines from single to double. The user can indent or number paragraphs, or add borders and shading. Paragraph formatting is applied to an entire paragraph. Paragraph alignment determines how the lines in a paragraph appear in relation to the left and right margins. The margin is the blank space between the edge of the paper and where the text is. For paragraph alignment the user needs to select the paragraph first and then click on the appropriate alignment button on the ribbon. Apart from clicking on the buttons, the alignment can also be changed by selecting the text and using the keyboard shortcuts as Ctrl+L = Left Align; Ctrl+R = Right Align; Ctrl+E = Center; Ctrl+J = Justify.

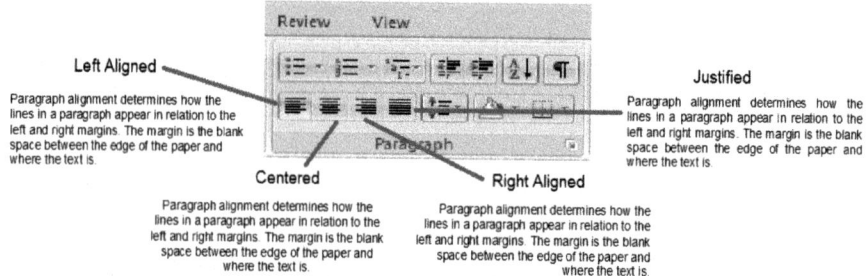

Left Aligned

Paragraph alignment determines how the lines in a paragraph appear in relation to the left and right margins. The margin is the blank space between the edge of the paper and where the text is.

Justified

Paragraph alignment determines how the lines in a paragraph appear in relation to the left and right margins. The margin is the blank space between the edge of the paper and where the text is.

Centered

Paragraph alignment determines how the lines in a paragraph appear in relation to the left and right margins. The margin is the blank space between the edge of the paper and where the text is.

Right Aligned

Paragraph alignment determines how the lines in a paragraph appear in relation to the left and right margins. The margin is the blank space between the edge of the paper and where the text is.

Line and Paragraph Spacing: Line spacing is defined as the vertical space between the two lines of the text in a paragraph. Some paragraphs may be single-spaced and some double-spaced. Single-spacing is Word's default setting. Line spacing can be increased as well as decreased. For changing the line spacing, the user needs to first select the paragraph, and then click on the button on the ribbon to change the line spacing as shown in the following figure.

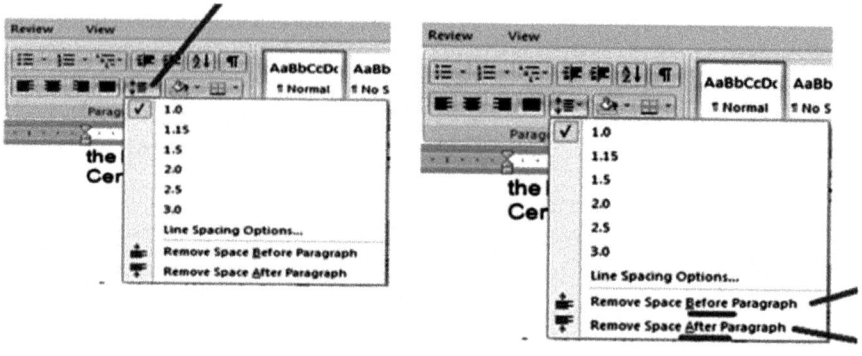

Paragraph spacing is the amount of space above or below a paragraph. Instead of pressing the enter key multiple times to increase space between paragraphs, the user can set a specific amount of space before or after paragraphs. For this option, the user needs to select the paragraph first and then needs to click on the button on the ribbon as shown in the following image.

Changing the background color of the paragraph: For this, the user needs to select the paragraph first and then click on the button at the ribbon and select the desired color. The background color of the paragraph will change as shown below:

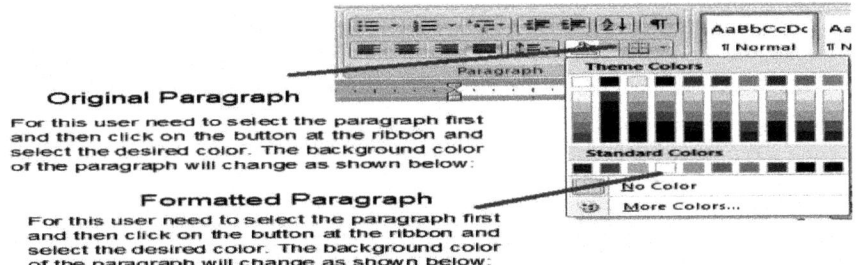

Changing paragraph indentation: An indent or indentation is the space between the left and right margin of a paragraph. To create an indent for the first line of text the user places the mouse pointer (cursor) to the beginning of the line and presses the tab key on the keyboard. A typical indent is five spaces from the left-hand or right-hand side of the page.

To change indentation the user needs to select the paragraph and then click on the increase or decrease indent button on the ribbon as shown here:

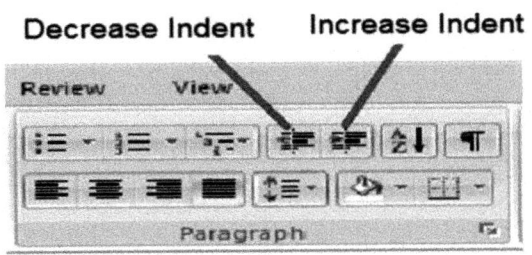

Hanging Indentation: Hanging indentation refers to a paragraph that has all lines indented except for the first one. This can be created by selecting all the lines of a paragraph and then increasing the indent of all the lines but for the first with the help of the ruler line as shown in the following:

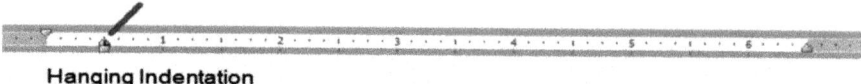

Hanging Indentation

Hanging indentation can be defined as a paragraph which has all lines indented but for the first one. This can be created by selecting all the lines of a paragraph and then increasing the indent of all the lines but for the first with the help of the ruler line as shown below:

Formatting Marks: These are defined as special marks that are not visible by default. The marks affect the display of the text in the document. The following are the marks that get displayed in the text:

Symbol	Name	Symbol	Name
¬	Conditional hyphen	——Page Break——	Pagination breaks
{.date.}	Field code	¶	Paragraph marks
↵	Line breaks	A.B	Space character
→	Tab		

The display of marks can be enabled by clicking on the mark button on the ribbon or by pressing Ctrl+Shift+8 key simultaneously. The text will look like this:

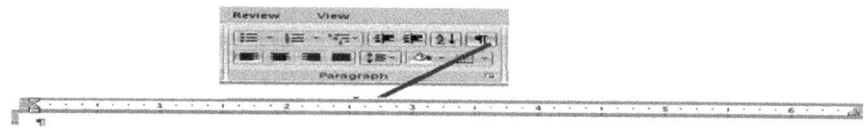

Hanging Indentation¶

→ Hanging·indentation·can·be·defined·as·a·paragraph·which·has·all· lines·indented·but·for·the·first·one.··This·can·be·created·by·selecting·all·the· lines·of·a·paragraph·and·then·increasing·the·indent·of·all·the·lines·but·for· the·first·with·the·help·of·the·ruler·line·as·shown·below:¶

Border and Shading: A user can add borders to pages, text, tables and table cells, graphic objects, and pictures. This option will be discussed under the "page layout" option.

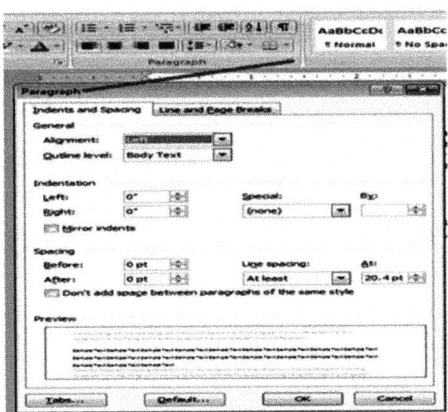

Paragraph Menu: All of the previous paragraph formatting commands can be performed with the help of the paragraph menu as shown in the next image:

Change Style Option: Users spend a lot of time formatting their documents to get them to look exactly as they want. There are several reasons to spend time styling the document. It may be because of publishing requirements of the document or just to give it a different look. While the styles gallery was introduced in Word 2007, in Word 2013 and later versions it has been made easier for the user to quickly change the look of entire document. In MS-Word 2007, the user was able to change the style of the document either at the home tab or by using the themes button on the page layout tab. But in Microsoft Word 2013, both are available in a unified design tab. This gives the user an option to change the style of the whole document in one click. For example, the user can change font type, font size, heading pattern, color, line-spacing indentation, and so on. Numbers of style patterns are already given and the user can also set their own style pattern. For implementing any style the user needs to select the entire document or the paragraph in which the style is to be implemented.

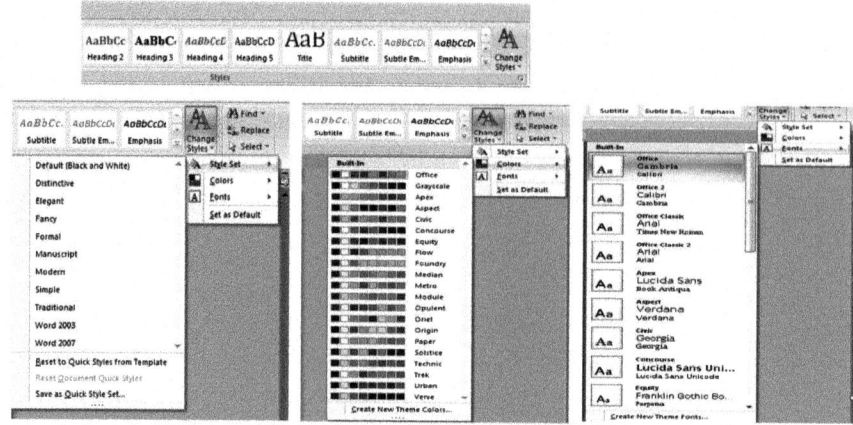

Use of Find, Replace, Go To, and Select Command (Editing Tab)

Word offers a unique feature that enables users to "find and replace" any specific content in the document such as words, images, captions, bookmarks, etc. On clicking on the "find" button on the ribbon, a window opens where user needs to write what they want to find in the "find what," column as shown in the following figure. The shortcut to execute this option is Ctrl+F.

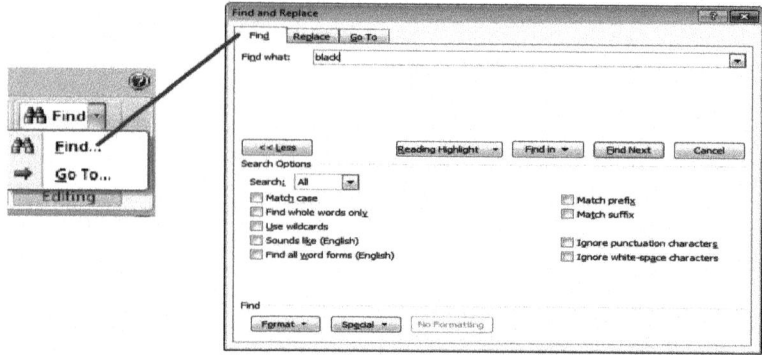

If the option "match case" is checked, the search will look for the exact case structure match of the word as entered by the user. If it is unchecked, Word will ignore case structure when looking for the word/words. Each time the word is found, the search will pause and the word found is then selected. Click on "find next" to continue searching or "cancel" to abandon the search. When the search reaches the end of the document, a window appears and tells the user that "Word has finished searching the document." Click OK to complete the process.

Replace: The "replace command allows the user to replace the word or group of words with another word or group of words. Use the keyboard shortcut by pressing the "Ctrl + H" keys simultaneously. Then the window below appears:

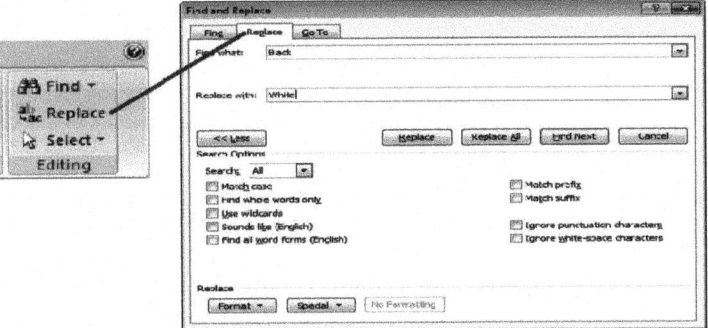

In the "find what" column, the user should type the word or the group of words that he would like to replace. Next, type the word or the group of words in the column called "replace with," then they click on the "find next" button. If the "match case" option is selected, the exact match will be searched in the document. Every time the desired word is found in the document, the search will stop and the word found will be highlighted. Next, the user clicks "replace" to replace the word in the "find what" field with the word in the "replace with" field or in "find next" to continue the search without replacing the word. The "replace all" button will replace all the highlighted (matched) words in the document without asking each time. The "cancel" button will abandon the search. When the search arrives at the end of the document or Word doesn't find the word the user is looking for, a window appears and tells the user that "Word has finished searching the document and has made n replacements," where n is the number of times the desired word was found and replaced. Click "OK" to complete the process.

Go To: If it is a large document, the user can use the "go to" command to go to a particular instance of content in the document, and he can also extend the search by using wildcards, codes, or regular expressions to find words or phrases that contain specific characters or combinations of characters. "Go to" can be used to jump to a desired page, section, line, bookmark, comment, footnote, endnote, field, table, graphics, equation, object, or heading. The shortcut key combination to execute this option is "Ctrl+G." The command works as is shown in the following figure:

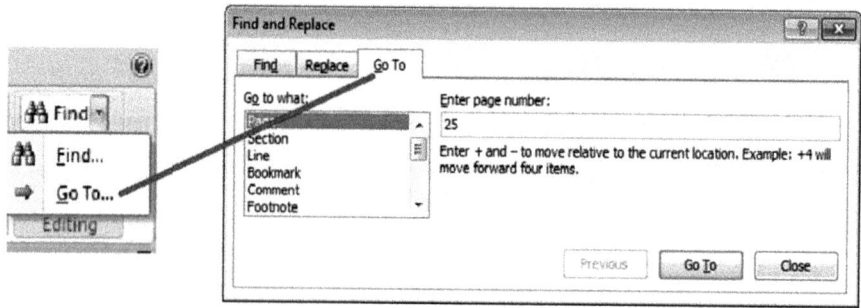

Select Option: The user has an option to select all text in a document by clicking on the select button on the ribbon and by clicking on the "select all" option (or alternatively by pressing the shortcut key combination of "Ctrl+A"), as shown in the following figure.

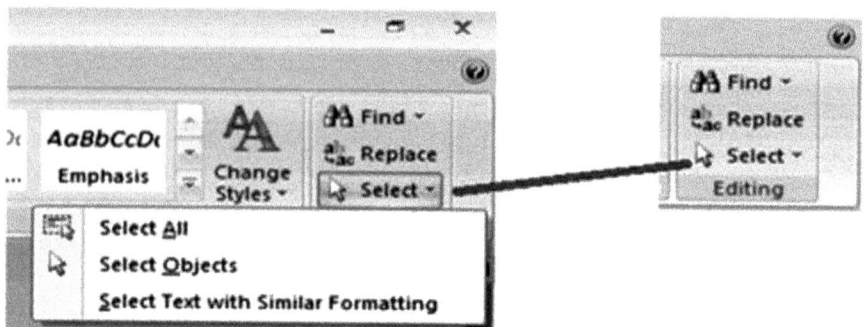

Another way of doing this is to select specific text or items by using the mouse or keyboard. The user can also select text or items that are in different places in the document. For example, the user can select a paragraph on one page and a sentence on a different page. To do this, the user needs to place the mouse pointer (cursor) in front of the first letter of the word, sentence, or paragraph to be selected. They then click the left button of the mouse and drag it to the point to which the text needs to be selected. To select a single word, the user places the mouse pointer anywhere on the word and just double clicks. The word will then be selected.

Alternatively, the same act can be done with the help of the keyboard. For this, the user needs to place the cursor at the beginning of the word and press "Shift+ Right Arrow" to select the line, press "Shift+ Down Arrow" ("Shift+ Up Arrow" will work in opposite direction), and to select a paragraph place the cursor at the beginning of the paragraph and press "Ctrl+Shift+Down Arrow" simultaneously.

3.2.4 Insert Tab

Cover Page: Under the "insert" tab, the cover page button offers a number of predesigned cover pages. The user can select a cover page and replace the sample text with their own. The cover page can be inserted by clicking on the cover page button and selecting the correct one. If users insert another cover page in the document, the new cover page will replace the first cover page. To delete a cover page, click on the "insert" tab, click "cover pages" and then click "remove current cover page."

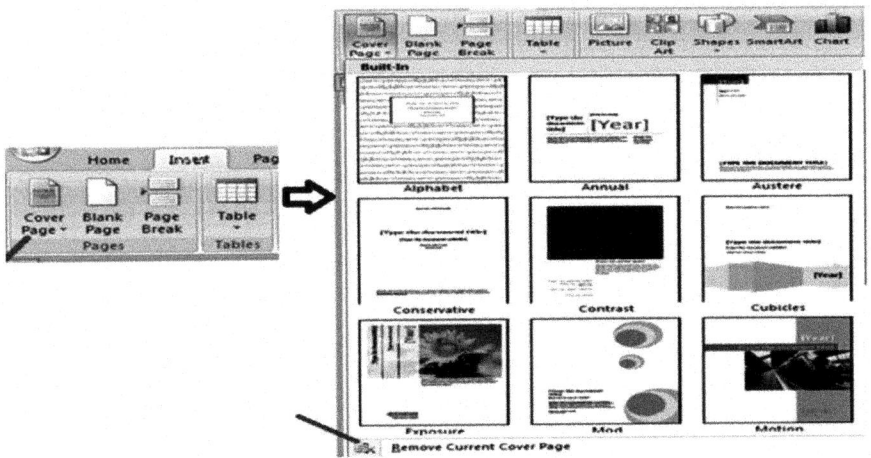

Blank Page: Click on the "insert" tab. Next click on the "blank page" button. A blank page will be inserted at the cursor position.

Page Break: Page break refers to the point where one page ends and a new page starts. Page break is automatically inserted at the end of each page when a user creates a document. However, Word gives an option of inserting a page break at the desired position to the user. The user should place the cursor or mouse pointer at the desired position, click on "insert" and then click on "page break."

To avoid page breaks from appearing in incorrect places (e.g., between lines of text), the user can adjust the page break settings for selected paragraphs. This can be done by performing the following steps:

1. Select the paragraph where the page break is to be inserted.

2. Click on the "home" tab and select the paragraph option. In the paragraph option click on the "line and page break" tab. The following options will be available:

    *   "Widow/Orphan control" places at least two lines of a paragraph at the top or bottom of a page.

    *   "Keep with next" prevents breaks between paragraphs you want to stay together.

    *   "Keep lines together" prevent page breaks in the middle of paragraphs.

    *   "Page break before" adds a page break before a specific paragraph.

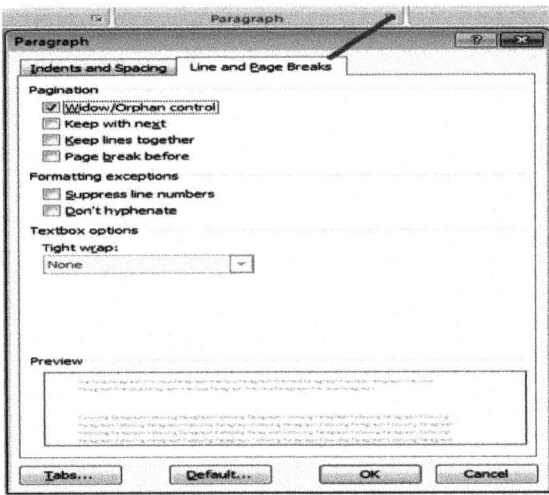

Table: A table presents data in rows and columns format. Microsoft Word provides the user with a very convenient option of creating a table that includes the desired number of rows and columns. To do this, the user needs to take following steps:

Select the "insert" tab and click on the "table" button. A drop-down menu will appear and gives the user an option of selecting the number of rows and columns with the help of the mouse as shown in the following figure:

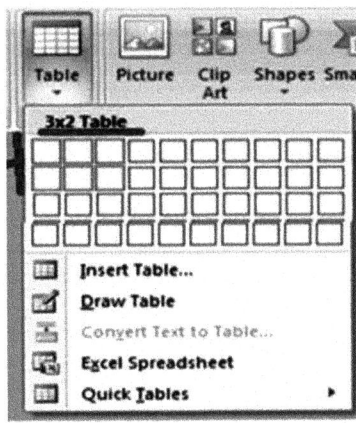

The preceding screenshot shows that the user is creating a 3 × 2 table, which means it consists of three columns and two rows. The table will appear at the mouse pointer position. As soon as the user presses the tab key in the last column of the table, a new row will automatically be added. The same can be done by clicking on the "insert table" option as show in the following figure. Here, the user needs to specify the number of rows and columns they want in the table.

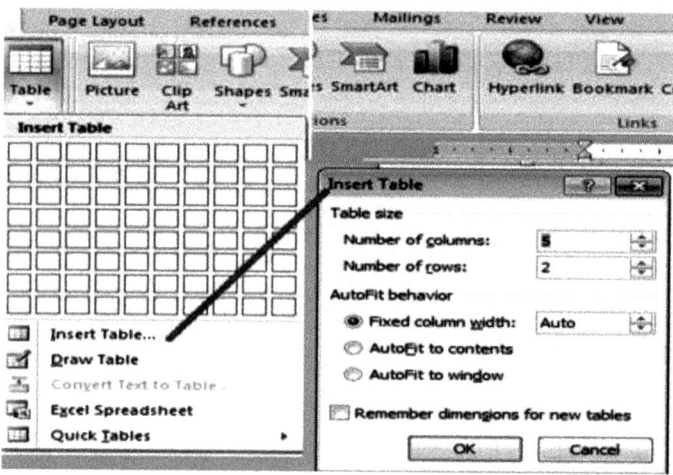

"Insert table" gives the following additional options to the user under the "autofit" option:

(*a*) **Fixed Column Width:** Here the user can specify the column width in inches.

(*b*) **Autofit to contents:** The column width will continue to increase as per the text entered by the user in the column window.

S. No	Name of the candidate	class		

(*c*) **Autofit to Window:** Here the column width will not change, but the text will shift to the next line of the column from first to second to third and so on.

Name of the Candidate	Class		

The dimensions of the previous table will remembered and the user need not specify them again.

Draw Table: By selecting the "draw table" option, the user can create a table using desired height and width of the rows and columns. When this option is selected, the cursor acquires the shape of a pencil with which the user can draw rows and columns.

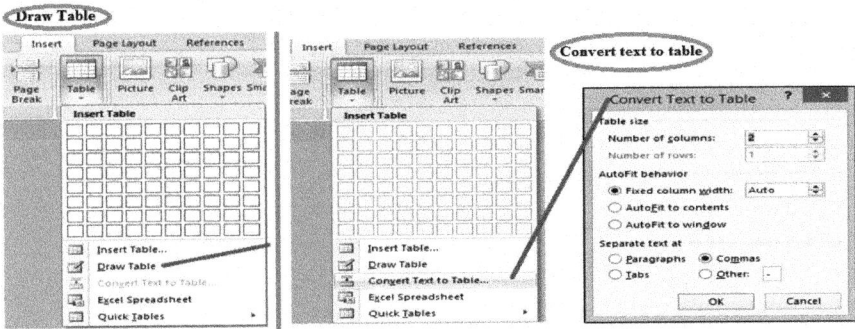

Convert Text to Table: This option lets the user convert the paragraph into the table, splitting the text into columns at each comma, full stop, or other characters specified by the user. For this, the user needs to select the text first, and only then will the option become active. Next, the user clicks on "convert text to table." The number of rows will automatically be determined by line breaks, so for example, if a block of text is divided by four line breaks, the table will have four rows. Columns are determined by commas, tabs, paragraph breaks, or another symbols the user has assigned. The result is shown as follows:

This option lets the user convert the paragraph into the table	splitting the text into columns at each comma	full stop	or other character specified by the user. For this user need to select the text first only then the option will become active	now click on "convert text to table."

Excel Spreadsheet: The user can actually insert an Excel sheet into the Word document by using this option. They can enter new data into rows and columns or they can copy data from an existing sheet. Once data is entered into the rows and columns of the spreadsheet, in a Word document it will appear similar to a table although it is technically known as a "workbook object."

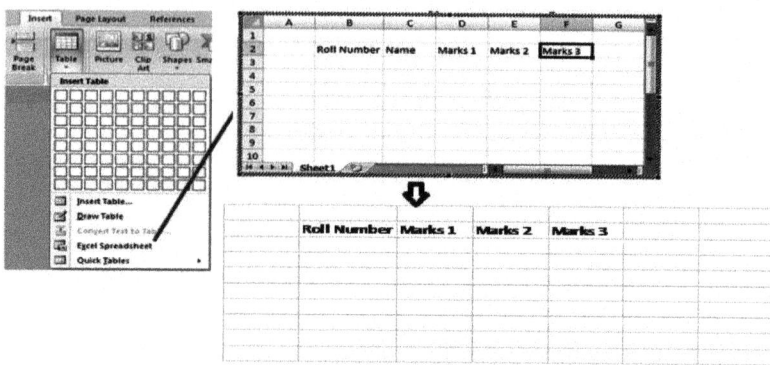

One can right click on the table created to change the borders and shading as show here:

Rol Number	Mark 1	Maks2	Maks3		

Quick Tables: Quick tables lets the user insert a quick calendar, matrix, or a tabular list and allows the editing of it. The user can create their own format and can save their work by clicking on the option "save selection to quick tables gallery."

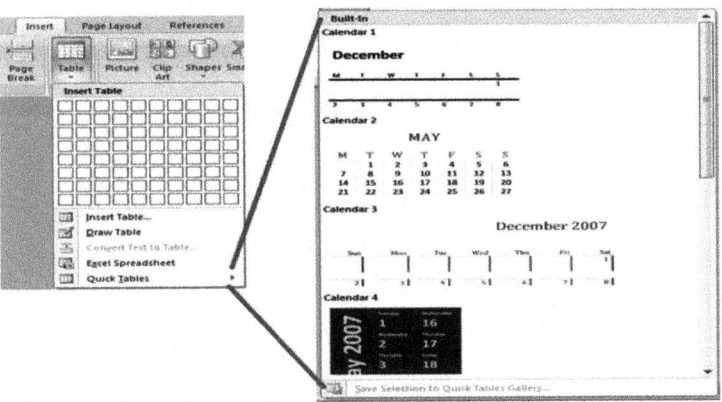

Formatting Tables: Select the table by dragging the mouse pointer from start till the end of the table or by left clicking on the sign shown in the circle in the following figure. Below the sign, a table can be selected.

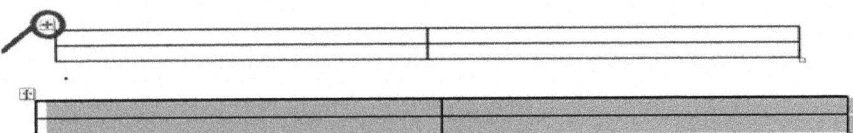

Once the user has left clicked on the sign inside the red circle a new tab named "table tools" appears on the ribbon. "Table tools" is divided into two tabs. "Layout" lets you add and remove columns, adjust height and width, and adjust text alignment. Many of these controls can be accessed directly from the right-click context menu. The "design" tab by contrast is all about how the table appears.

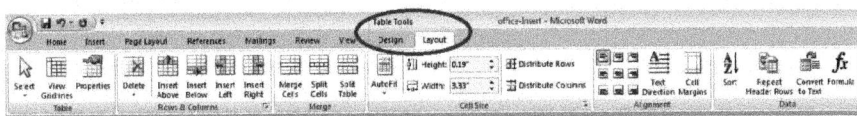

Alternatively, a right click on the sign in the circle will yield a drop-down menu that gives all the options for formatting the table as shown in the following:

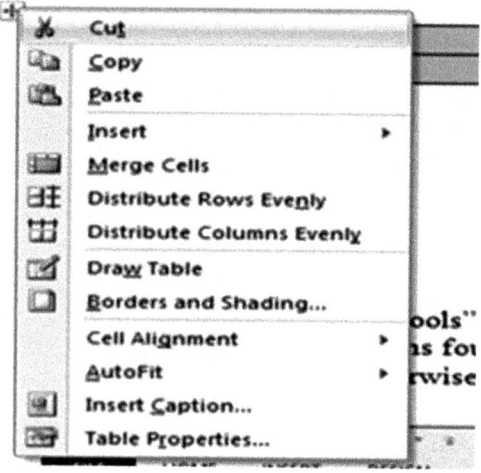

In the above drop-down menu, clicking on the "cut" option will remove the table along with its data. The "copy" option will make a copy of the table, and the "paste" option will paste the table at the position of the mouse pointer with full data. "Paste as nested table" will append the cells into the already existing table provided the mouse pointer is placed at the starting point of the row just after the last column of the table.

Insert Option lets the user insert new row, column, or cell in the table.

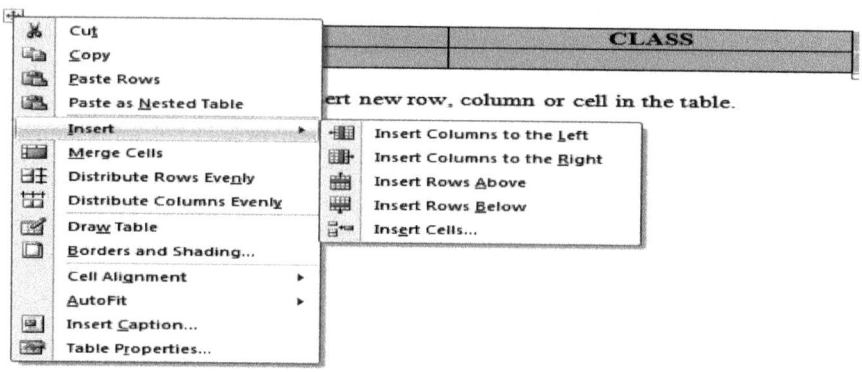

button where a drop-down menu will appear with the "text direction" option. The user should click on the button multiple times to set the direction of the text as desired. The result is shown in the following figure:

Original Table

Name	Class	Marks
Anjali	XI	80
Akshat	XI	87
Bhanu	XI	90

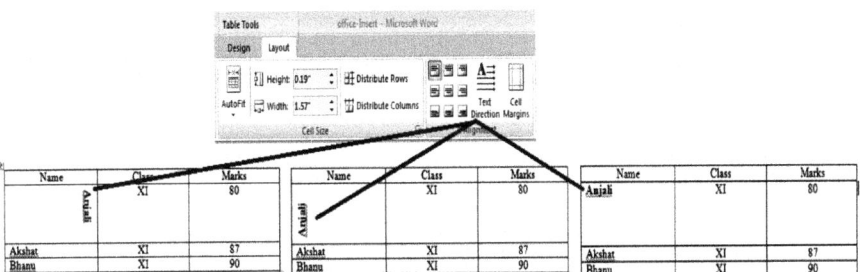

Alternatively, the same can be done by placing the mouse pointer in the desired cell, clicking on the right mouse button, and left clicking on the text direction option, after which a new window will pop up and the user can select the desired direction of the text, Clicking on the OK button gives the result as shown in the preceding figure:

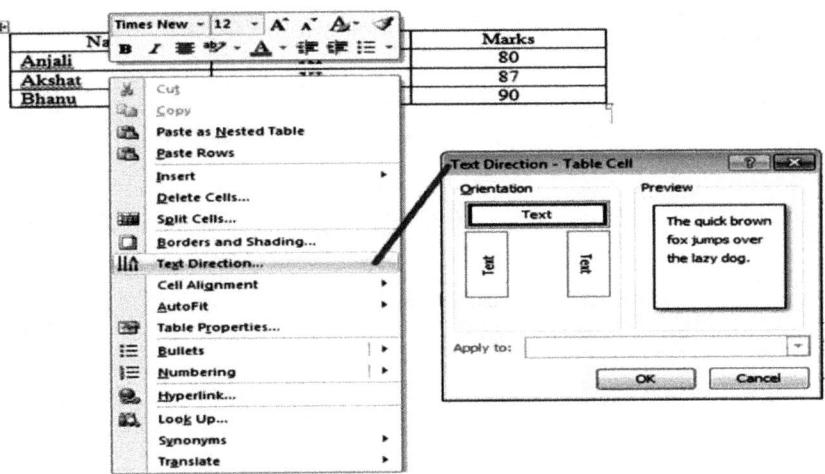

Cell Margins

The user can set cell margins as per need, and can also set the spacing between cells. Select the table or place the mouse pointer inside any cell. Select the "cell margin" button on the layout tab. A new window will open that will allow the user to set top, bottom, right, and left margins. Selecting the option "allow spacing between cells" sets the required space among all the cells as shown here: **Original Table**

Name	Class	Marks
Anjali	XI	80
Akshat	XI	87
Bhanu	XI	90

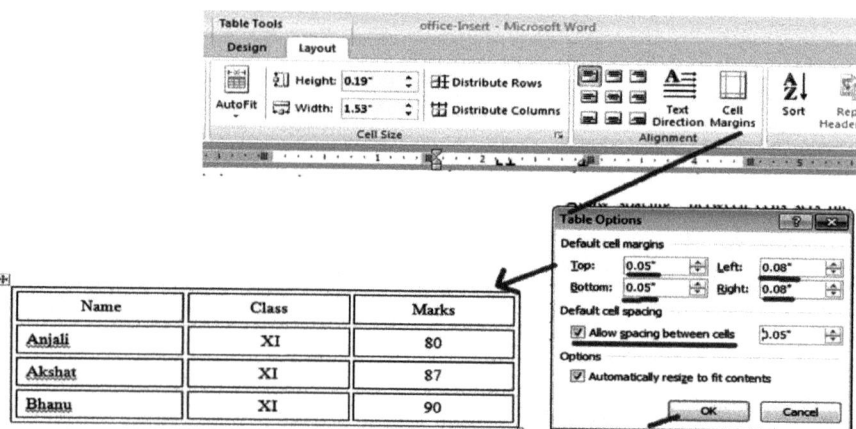

Insert Caption: A caption is a heading given to a figure, equation, table, or any other object. A caption is a numbered label, such as "Figure 1," "Table 1," and so on. It's comprised of text such as "Figure," "Table," "Equation," or something else followed by an ordered number or letter such as "1, 2, 3 ..." or "a, b, c ..." and so on. To apply a caption, the user selects the table or figure, clicks on the right mouse button, and clicks on the "insert caption" button on the menu. Another window will open with certain options. The user selects the option from equation, figure, or table in label column. The position column allows the user to place the caption either above or below the figure, equation, or a table. The "new label" button allows the user to define their own caption. Checking on the "exclude caption" box hides the label and shows only the numbers.

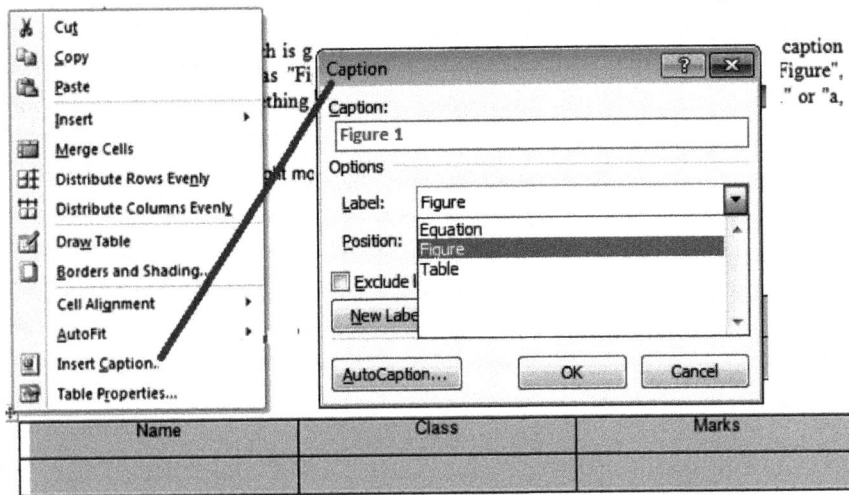

Selecting the "caption numbering" button opens another window that gives the user a number of formats for caption numbering as shown in the following figure:

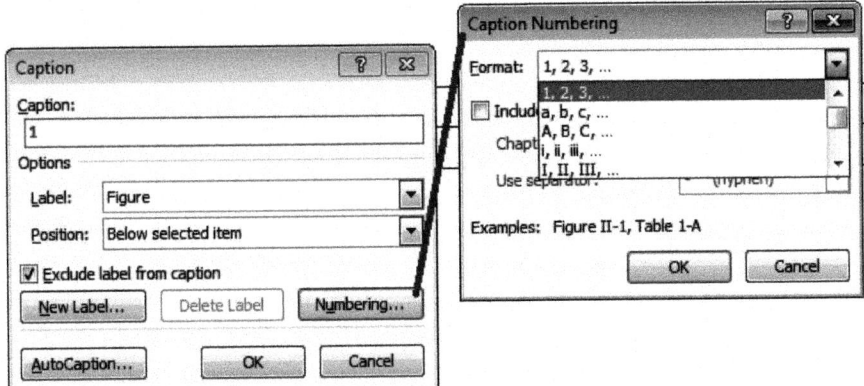

Table Properties

Select the table, click the right mouse button, and select the table property option. Another window will open that gives users an option to set the height, width, and other parameters of the row, column, cell or the entire table as shown in the next figure:

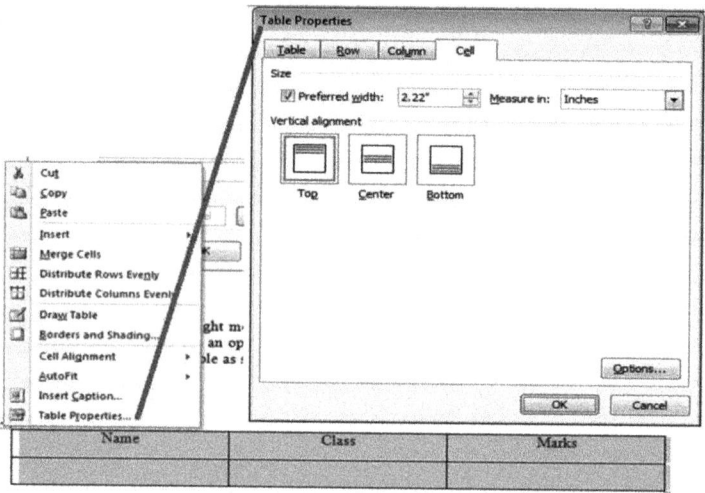

Sort Option

Name	Class	Marks
Nishant	XI	87
Bhanu	XI	90
Akshat	XI	80

Select the table or place the pointer in any column below the header row, click on the "sort" button in the "layout" tab. A new screen will pop up asking the user to specify the sort options. Data can be sorted either alphabetically or numerically in ascending or descending order as shown in the following figure:

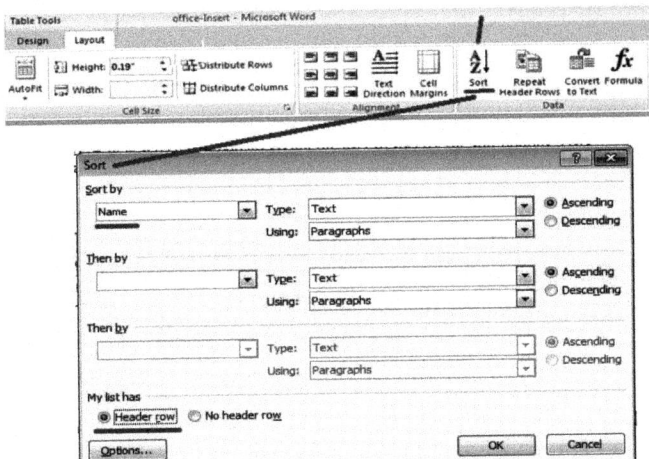

Here the sorting is done on the basis of name. If "no header row" is selected, then the first row of the table will also be taken for the sorting purpose, so the user needs to select header row to get the correct result. No header row has to be selected if there is no header row defined. Output of the sorting is shown in the following figure:

Original Table

Name	Class	Marks
Akshat	XI	87
Bhanu	XI	90
Nishant	XI	80

Sorted Table on Name Field

Name	Class	Marks
Akshat	XI	87
Bhanu	XI	90
Nishant	XI	80

Repeat Header Row: Selecting this option will insert header row in every page in case the table extends beyond one page. Command can be simply executed by selecting the header row first and then clicking on the repeat header row button in the layout tab as shown in the following figure:

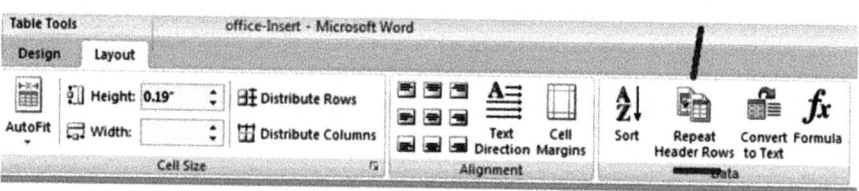

Convert to Text: This helps in converting the table into the text format. The user needs to choose which text character to use as a separator. The user selects the table, clicks on the "convert to text: button, and another window will appear and ask the user to define the separator. Defining it and clicking on "OK" will convert the table to text as shown in the following figure:

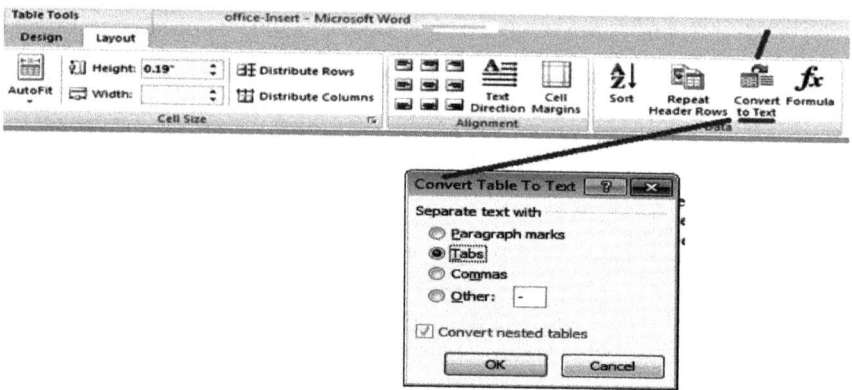

Formula: Numbers of formulas available to users are provided by Word. Word provides the user an option of performing calculations and logical comparison in a table with the help of formulas. Some formulas provided by word are sum, product, average, count, max, and min. The arguments to be used with these functions are left, right, above, and below. To insert the formula, place the mouse pointer in the blank cell of the table. A "layout tab" will appear on the ribbon. Click on layout tab and in it click on the "formula" button. A screen will appear as shown here:

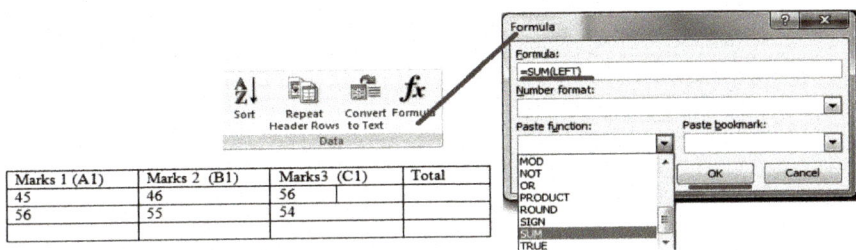

Marks 1 (A1)	Marks 2 (B1)	Marks3 (C1)	Total
45	46	56	
56	55	54	

The value in the cell can be changed as shown here:

=SUM(**ABOVE**)	adds the numbers in the column above the cell where formula is written
=SUM(**LEFT**)	adds the numbers in the row to the left of the cell where formula is written
=SUM(**BELOW**)	adds the numbers in the column below the cell where formula is written
=SUM(**RIGHT**)	adds the numbers in the row to the right of the cell where formula is written

If we want to work with other available functions, put the mouse pointer in the cell, click on the formula button, in place of the formula option after = sign type the function name from the list of available functions such as "average."

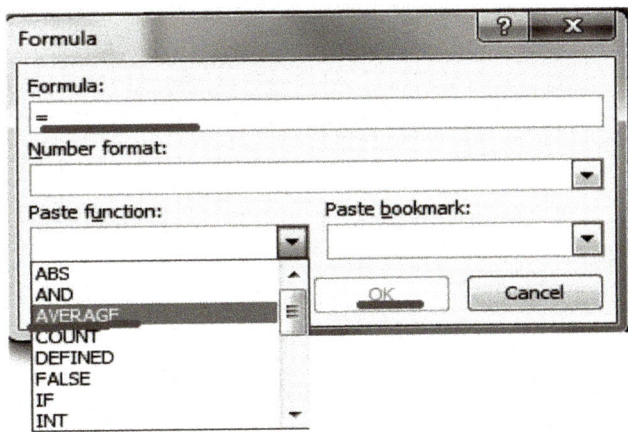

=AVERAGE(**ABOVE**)	will find the average of the numbers in the column above the cell where formula has been paste
=AVERAGE(**BELOW**)	will find the average of the numbers in the column below the cell where formula has been paste
=AVERAGE(**LEFT**)	will find the average of the numbers in the column which are to the left of the cell where formula has been paste
=AVERAGE(**RIGHT**)	will find the average of the numbers in the column which are right to the cell where formula has been paste

Clicking on "OK" will find the average. To multiply two numbers, click "product" and type the location of the table cells as right, left, below or above. If changes are made to the numbers that are added or multiplied, select the column where the formula is written and press F9 to bring the change in the results, or right click on the mouse and from the menu click on the "update field" option. The cell value will then be updated.

Illustrations

Picture: This option allows the user to insert a picture from a file in the active document. Place the mouse pointer at the location where you want to insert a picture. Click on the "insert" tab, and then click on "picture"

button. The picture library will open on the screen as shown in the following figure. Left click on the desired picture and it will appear in the document at the position of the pointer.

Clip Art: Clip art is defined as a collection of pictures or images that can be imported into a document. Clip art galleries contain hundreds of thousands of images. Clip art is organized into various categories, such as people, objects, nature, etc., and is especially helpful when browsing through thousands of images. Most clip art images also have keywords associated with them. For example, a picture of a teacher in a classroom may have the keywords "school," "teacher," "classroom," and "students" associated with it. Most clip art programs allows the user to search for images based on these keywords.

Step 1: Place the mouse pointer at the location where you want to insert an image from clip art. Click on the insert" table and then click on the "clip art" button.

Step 2: When the clip art gallery opens, select the desired category, and select the desired image and click on it. A clip art image will be inserted into the document.

Shapes: Shapes provides the user a number of shapes such as lines, circle, rectangle, square, arrows, and flowchart symbols. Place the pointer at the position where the shape is to be inserted. Click on the "shapes" button. When the drop-down menu appears, select the desired shape and click on the document. The shape will appear in the document.

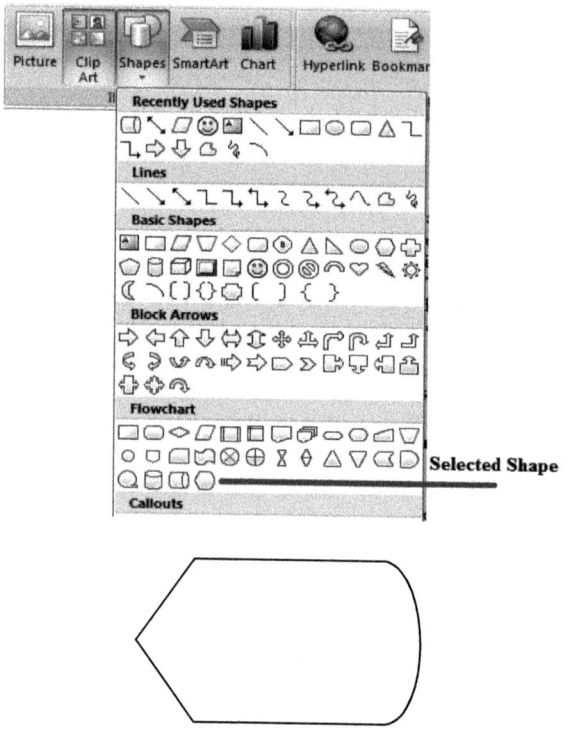

SmartArt: SmartArt is a way to represent text in a visually appealing manner. It is used to present information in a simple manner. SmartArt can be used to create an organizational chart, a decision tree, a pyramid or a matrix structure, illustrate steps in a process, or to display events in a timeline. Place the mouse pointer at the location where you want to insert SmartArt. Click on the "insert" tab and then click on the "SmartArt" button. A new window will appear as shown in the following figure.

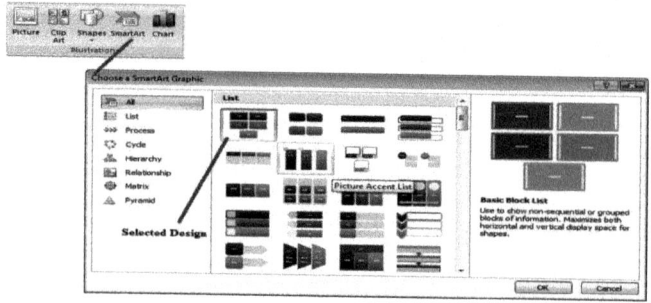

The user should select the desired image and click "OK." The shape will appear in the document and the user will be asked to input data. The output of this option is shown in the following image.

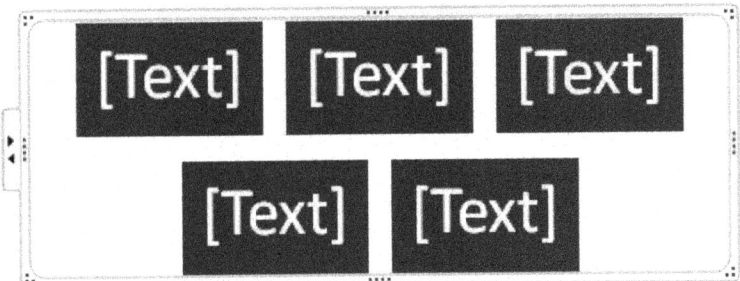

Chart: This option inserts a chart in the document at the desired location. The chart may be a bar, pie, line, area surface, etc. Place the mouse pointer at the location where you want to insert a chart. Click on the "insert" tab and then click on "chart" button. A new window will appear as shown in the following figure. Select the desired shape.

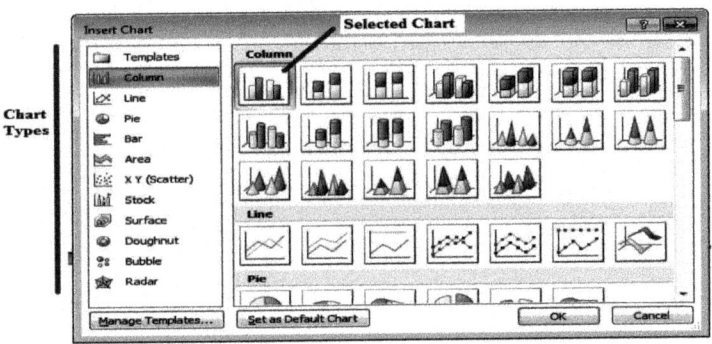

Click on OK. A chart will appear in the document.

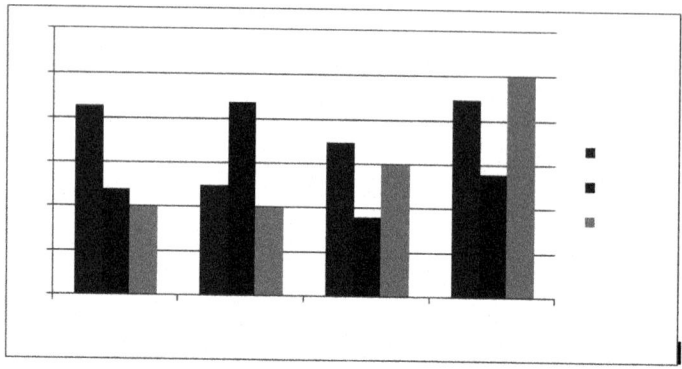

Links

The "Insert" toolbar provides the following options: "hyperlinks," "bookmarks," and "cross-references," all of which can be used as "links." Hyperlink allows the user to link pieces of text to locations on the computer, network, or the Internet. It does not have to be an Internet URL; it can simply refer to a location on the computer or another location in the same document. Click on the "insert" tab and click on the "hyperlink" Button. A window will appear as shown in the following figure:

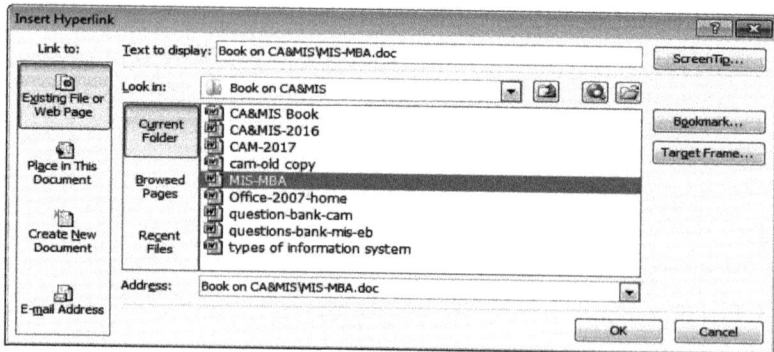

Clicking on OK will insert a hyperlink in the document. Press Ctrl + Left Click on the linked text, image, or document, and it will open on the screen:

Book on CA/MIS/MIS-MBA.doc

Alternatively, a user can also select text, right-click, and choose "hyperlink" from the context menu. In the "insert hyperlink" dialog, paste or type the address in the provided space. For example, creating a link to MIS notes: Select the word MIS and right click on it. A context menu will appear; select hyperlink and a window will appear to define the full path. Now when pressing OK a hyperlink will be created which will be activated on pressing Ctrl + Left Click simultaneously.

Bookmark: When a user is working on a big document they need to go up and down the document again and again. To facilitate the quick movement across the document, a user can create "bookmarks." A bookmark allows the user to assign names to text or to positions in the document. With its help the required para, text can be located easily. Different documents can be assigned same bookmark names. To define a bookmark, the following steps are to be taken:

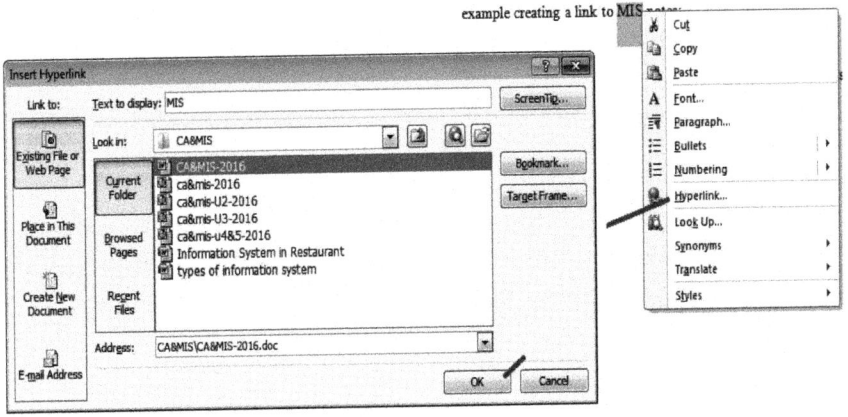

(*a*) Place mouse pointer at a place where the bookmark is to be inserted.

(*b*) Click on the bookmark button on the insert tab. A menu will appear as shown here:

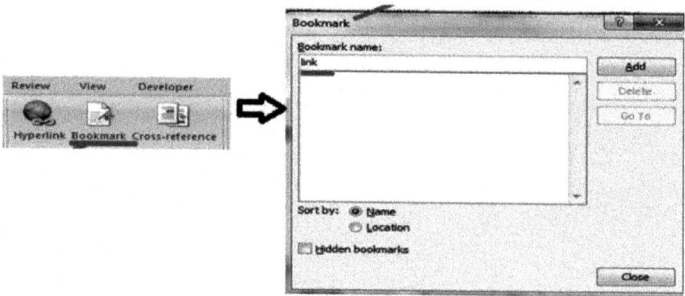

(*c*) Enter the name of the bookmark and click on "add." The bookmark will be created and the window will close down.

To go to the defined bookmark, use the "go to" option from the home tab. While defining the name of the bookmarks, users need to take care of the following:

(*a*) A bookmark name must start with a letter.

(*b*) A bookmark name cannot contain spaces or punctuations marks.

One precaution that needs to be taken is if the text is not selected when a bookmark is created and then the text has moved to some other location, the bookmark will stay where it was created. So to move the bookmark along with the text, the user needs to select the text before creating a bookmark.

Cross-reference: Cross-reference works the same as the bookmark. The difference is that the cross-reference creates a hyperlink to move to the referenced text, paragraph, page, bookmark, or heading, whereas in bookmark, the user needs to use the "go to" command.

Working with Header, Footer, and Page Numbers: Headers and footers are used to repeat the same information at the top and/or bottom of each page. For example, when the user wants to have the title of the document on each page and the name of the document creator on the bottom of the page. The line appearing at the top of each page is called the header, and the line appearing at the bottom of the page is called the footer. Similarly when a user wants to number the pages of the document the page number command is used. The working of the preceding three commands can be described as follows:

1. Click on the header button and a drop-down menu will appear with various header styles. Select the one you want to use.

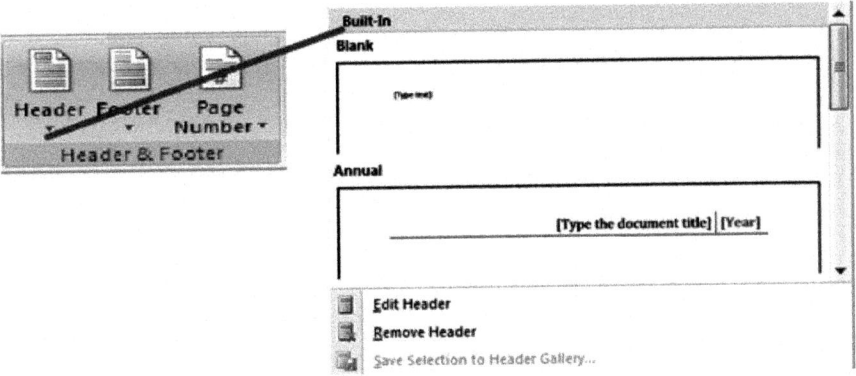

2. After you select the desired style, type the header text in the document, click on close header/footer button, and the header will appear in the document. Next click on the footer option, type the footer text in the document, and click on close header/footer button. The header and footer will appear on each page of the document as shown here:

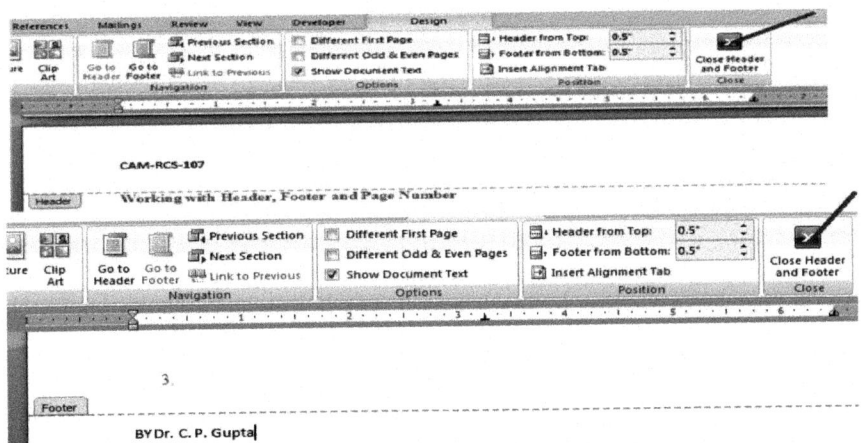

In the similar manner, the page number can be added in the document. The user can place it anywhere within a header or footer as shown here:

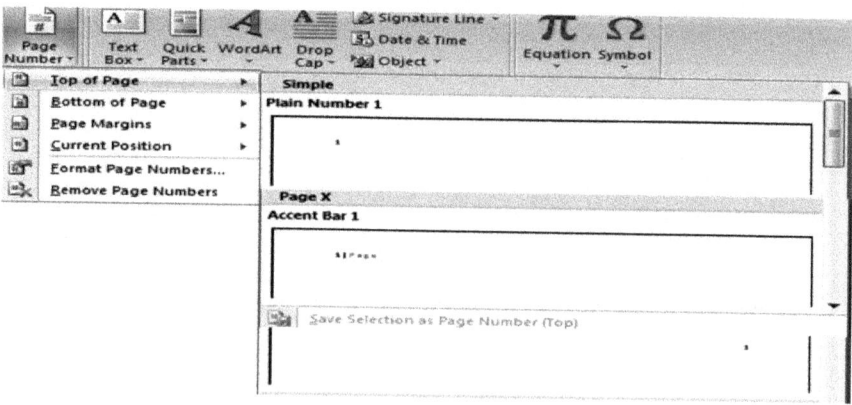

If you want to format page numbers, click on "page number" from a drop-down menu, and click on "format page number." A dialog box will open that will allow the user to change the number format, add chapter numbers, and dictate where pages start.

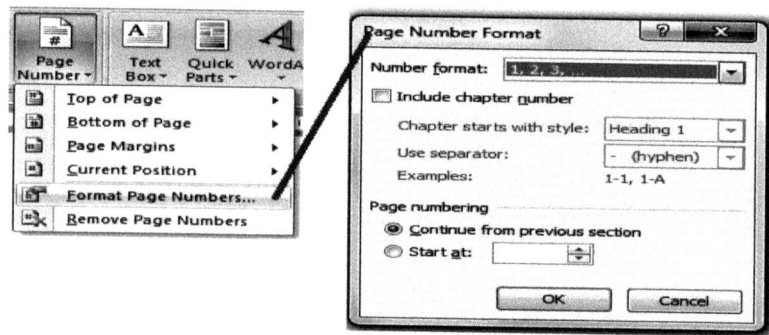

Text Box

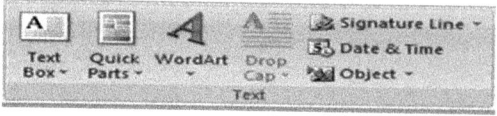

A text box gives a user freedom to type text and position the box anywhere in the document by dragging it with mouse. For inserting the text box, the following steps need to be taken:

(*a*) Select the insert tab and click on "text box" button. A drop-down menu will appear with a number of text box styles. Select the preferred one.

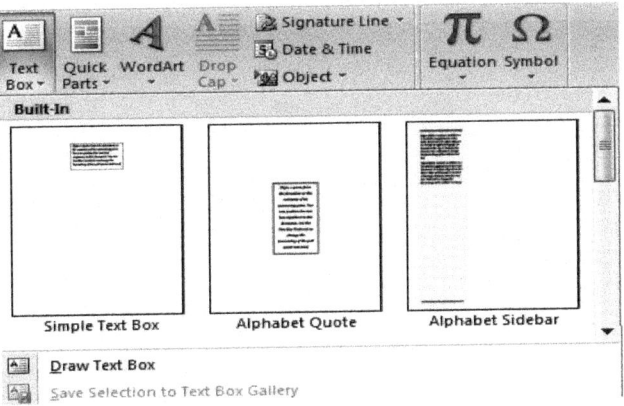

(*b*) Clicking on the selected style will make that box appear in the document at the mouse position. Clicking inside the box will allow the user to type the required text in the box.

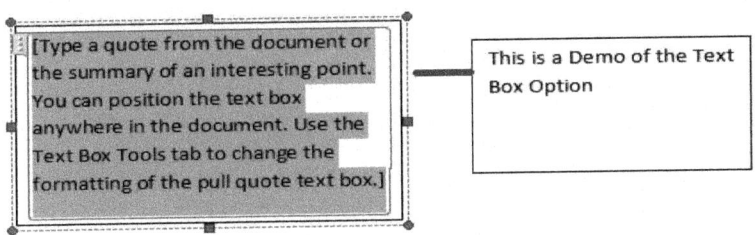

(*c*) On double clicking on the boundary of the text box a new "format" tab appears on the ribbon. This gives users the option to format the text and the box in number of ways:

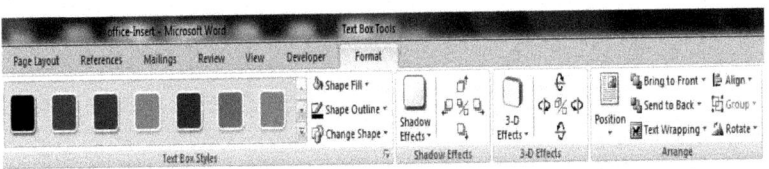

Date and Time: This gives users a number of formats in which date and time can be inserted at the mouse pointer position as shown here:

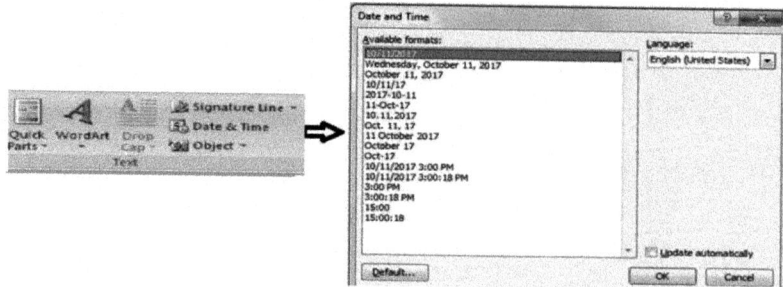

Object: This option lets the user insert into the active document an object such as a PDF file, a bitmap image, a graph, or even a document file. Place the mouse pointer at the position where an object is to be inserted. Click on the "object" button, and a drop-down menu will appear giving two options: "object" and "text from file" as shown here:

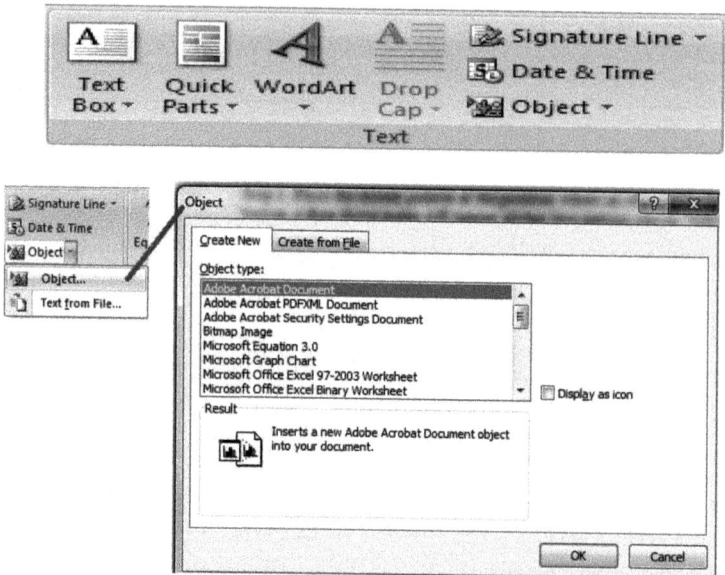

Clicking on "object" will open a dialogue box that gives options of various object types. The user should select the desired option and click on ok. The object will get inserted in the active document. The "text from file" option inserts the selected file into the active document.

Equation and Symbols: This lets the user enter various mathematical and algebraic equations into the active document, that otherwise cannot be typed as shown here:

$$(x + a)^n = \sum_{k=0}^{n} \binom{n}{k} x^k a^{n-k}$$

$$f(x) = a_0 + \sum_{n=1}^{\infty} \left(a_n \cos \frac{n\pi x}{L} + b_n \sin \frac{n\pi x}{L} \right)$$

Symbols: Lets the user insert various symbols that are not available on a keyboard.

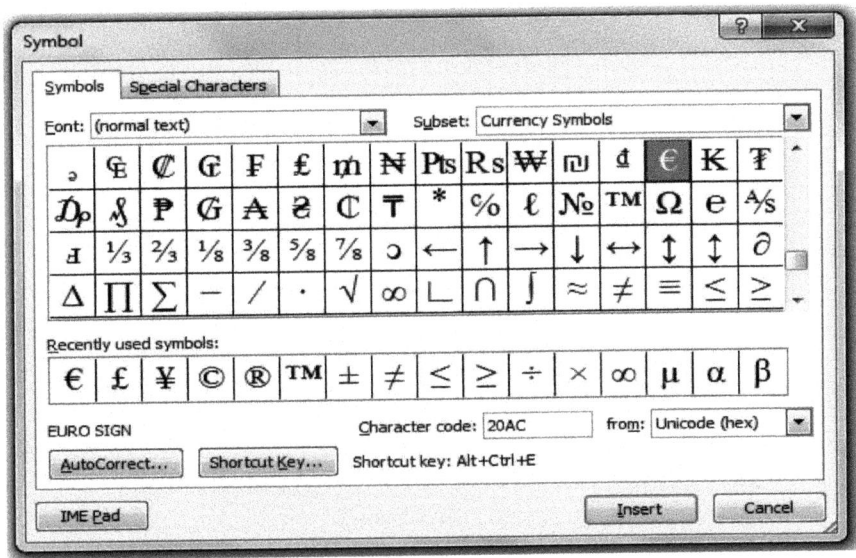

WordArt: WordArt is a utility provided by MS Word that lets the user insert decorative text into the document. It provides special effects like outlines, gradient glow, shadow, bevel, textures, and 3-D effects to the text. These effects are not otherwise available in the font option of the home tab. WordArt provides many preset styles that can be modified by clicking on the inserted WordArt text that activates the "format" tab on the ribbon. For working with WordArt, the user needs to take the following steps:

1. Select the "insert" tab and click on the "WordArt" button. A drop-down menu will appear with preset styles.

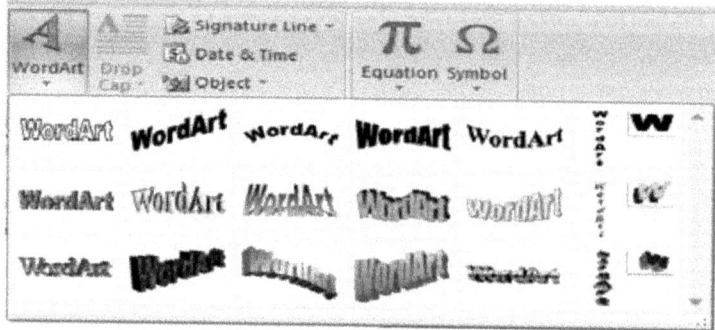

2. Click on any style of your choice and a dialogue box will open where the user will be asked to enter the text.

3. After entering the text, the user clicks on "OK" and the text will get inserted into the document at the mouse-pointer position.

4. Clicking on "demo" will open the "format" tab on the ribbon which allows the user to format the image created.

AutoCorrect: The AutoCorrect feature is provided by MS Word to correct typos, capitalization errors, and misspelled words, as well as automatically inserting symbols and other pieces of text. By default, AutoCorrect uses a standard list of typical misspellings and symbols, but the user can modify the entries in this list. AutoCorrect options can be set as follows: Click on the window button, then click on the proofing option. This will open a window where a user can set the AutoCorrect options as shown in the following figure:

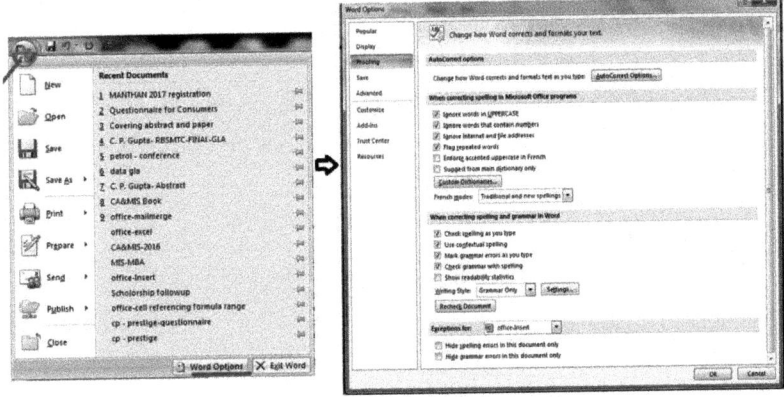

Spell Check: Spell check is a software program that corrects spelling errors in word processing, e-mail, and online discussions. Spell check identifies and corrects misspelled words. In Microsoft Word, spell check options such as spelling and grammar are found in the "proofing" window as shown in the preceding figure and under the "review" tab as shown in the following figure:

Step 1: Click on the "spelling and grammar" button under the "review" tab.

Step 2: The following window will pop up and Word will start suggesting the corrections to spelling and grammar as shown in the following figure:

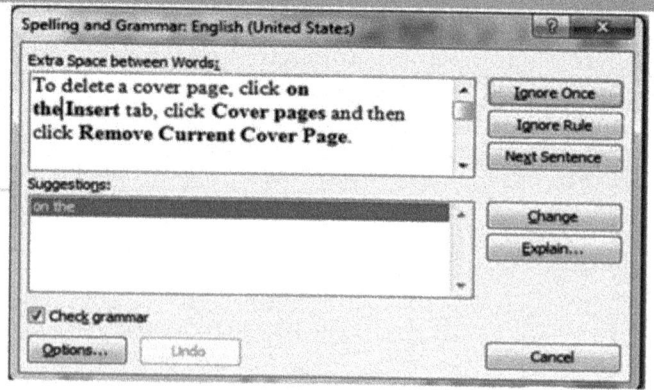

3.2.5 The Page Layout Tab

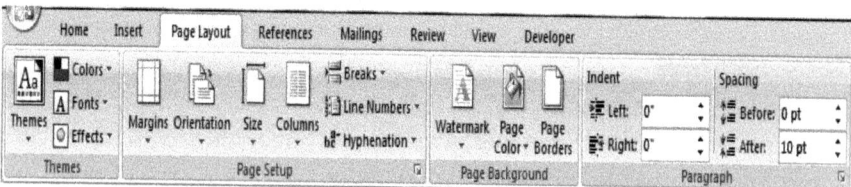

The page layout tab in MS-Word gives users an option to set the dimensions of a page in the document. IT sets the way a page will appear when it is printed. The page layout tab helps users set the margins, page orientation, size, number of columns, page color, and a number of other options. Users can apply a preset design to the document by using the available themes and color schemes.

Theme Button

Themes are preset formats of the document that allows users to change the way a document looks in terms of combination of colors, font styles, and formatting effects. For applying theme to the document, users need to apply style (found under the home tab) option first. The following steps are to be taken:

Step 1: Click on "home" tab, then click on any available style of your choice.

Step 2: Click on the "page layout" tab and select the theme of your choice. On the basis of the selected theme, the color, fonts, and effects will appear. For example, let us select the theme "flow." All three options of "colors," "fonts," and "effects" will appear as shown in the following screen-shot:

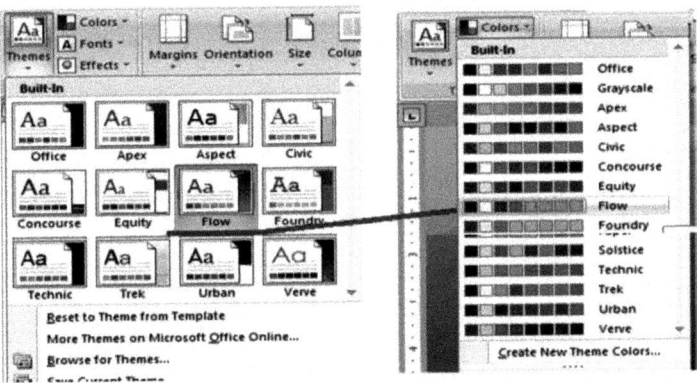

Step 3: The fonts option lets the user change the predefined font pattern in the "style" option that is already selected.

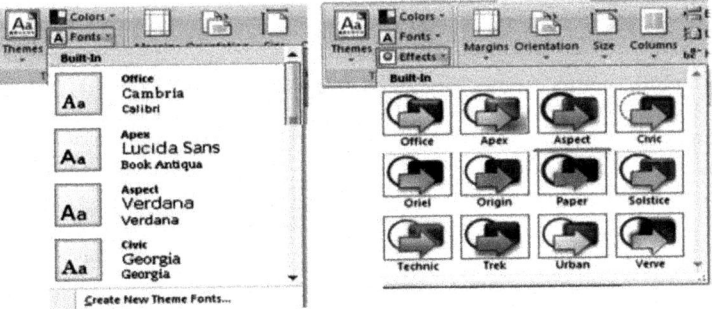

The Page Setup Option in Page Layout Tab

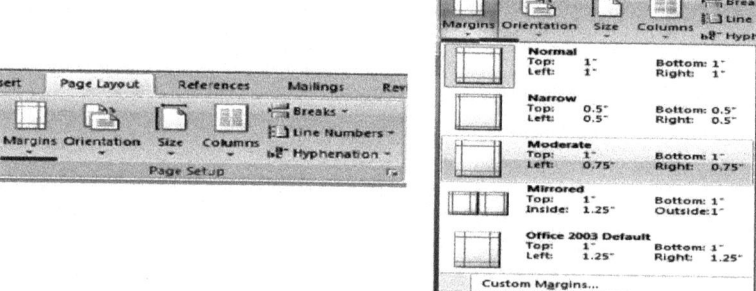

Margins: This option provides the user with preset options of the left, right, top, and bottom margins of the document. Users can click on any available option and the settings will be applied to the entire document. The custom margins option opens a dialogue box that allows the user to manually change the margins setting of the document. Orientation, page size, and page layout can also be changed. This can also be done by clicking on "page setup" as shown in the following below:

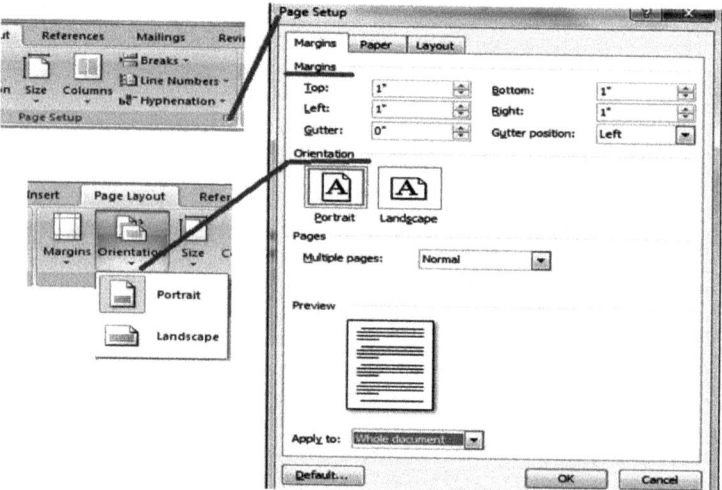

Page Orientation: There are two types of page orientations. One is portrait and the other is landscape. In *portrait* mode the image displayed on the paper is taller than it is wide, and in *landscape* mode the image displayed on the paper is wider than it is tall. By default the mode is set to portrait, but with the "page setup" option it can be changed to landscape.

This option can be executed in two ways: by clicking on the "orientation" button on the page layout tab, or by clicking on the "page setup" option as shown in the preceding figure.

Paper size: Papers are available in number of sizes such as letter, A3, A4, and so on. Users can set the paper size as per need and type of paper used. This can be done in two ways. One is by clicking on the "size" button on page layout tab, and the second is by clicking on the "page setup" option and then selecting "paper" as shown in the following figure:

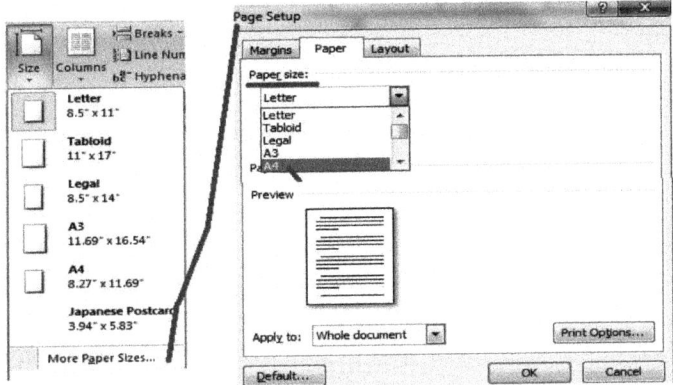

Clicking on the "more paper sizes" option under the "size" button also opens the same dialogue box as clicking on the "page setup" option.

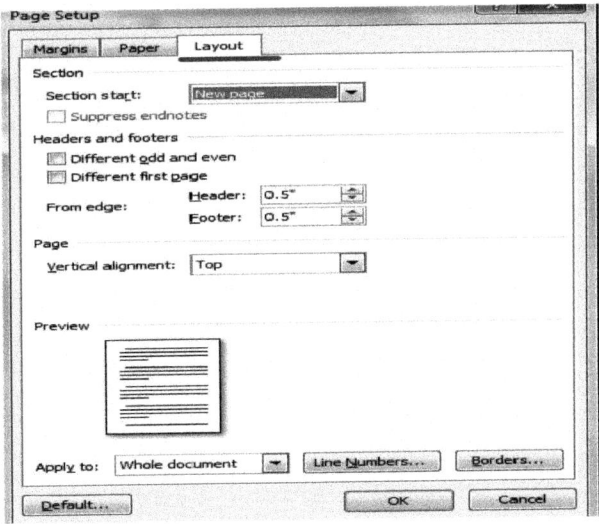

3.2.6 Mail Merge

Users need to communicate using many entities in day-to-day work, be it by letters, declarations, invitations, mail, invoices, etc. MS-Word provides a feature called Mail-Merge that helps the organization generate effective written communication. This feature of Word helps in sending letters, printing addresses on envelops, creating labels, and sending invitations by merging the main text with the available database (aka mailing list). The new document is called a Mail-Merge document.

Mail merge process: Process of mail merge creates three documents defined below:

- **Main Document:** This document contains text that is to be sent to all the members of the mailing list.

- **Mailing List:** This document contains all details of the members to whom the mail is to be sent. For example, it might contain all member addresses that are to be printed on the envelopes or the label.

- **Merged document:** A new document is created and is a combination of the main document and the mailing list.

What documents can be created?

- letters
- e-mail
- envelopes
- labels

The following steps are to be taken to create a mail-merge document:

Step 1: Click on the "mailing" tab on the ribbon, and the following options will appear as shown in the following figure. If the mailing list already exists, then users can continue with the drafting of the letter. Otherwise, users need to create a mailing list by clicking on the "select recipients" button. The following three options will appear as shown in the following figure:

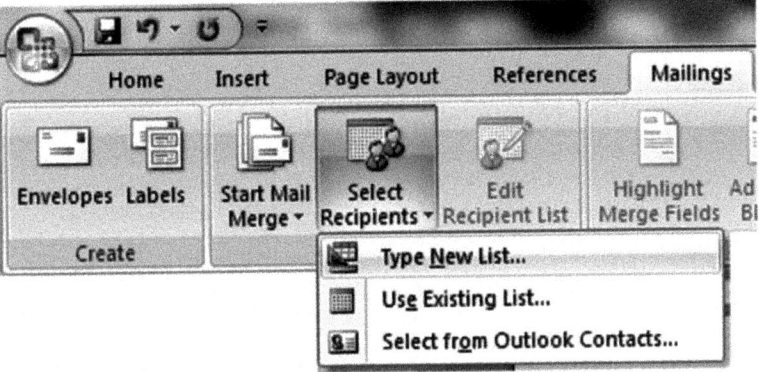

If there already is a list, users need to select "use existing list." To use the Outlook address book, users need to choose "select from Outlook contacts," otherwise click on "type new list." Clicking on "type new list" will open another dialogue box that lets users create a new mailing list when they define the structure and enter the values. This box gives users an option of customizing the values that they want to take use the structure and for this they need to click on the "customize column" button which allows them to add, remove, or rename the fields provided by default as per the need. The "customize address list" window also allows users to change the position of the fields provided by default. This can be done by selecting the field and clicking on the "move up" or "move down" option. The process of doing this is shown in the following figure:

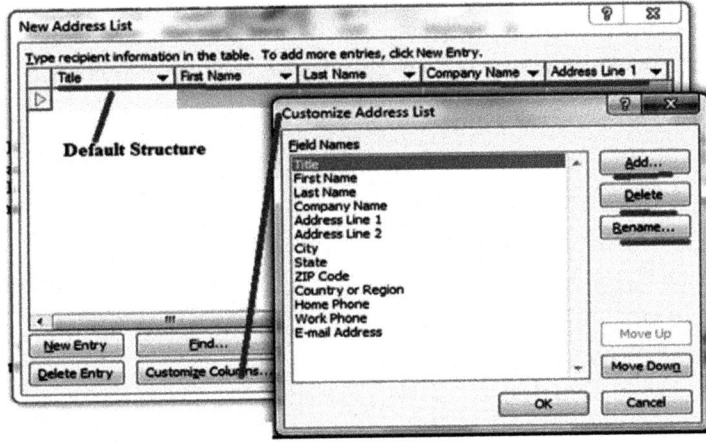

Here's an example to rename the "ZIP Code" as "Pin Code" and remove the field E-mail Address. Select the field "Zip Code" by placing mouse pointer on it, then click on "rename" button. A pop-up will appear asking the user to change the name of the field as shown in the following figure:

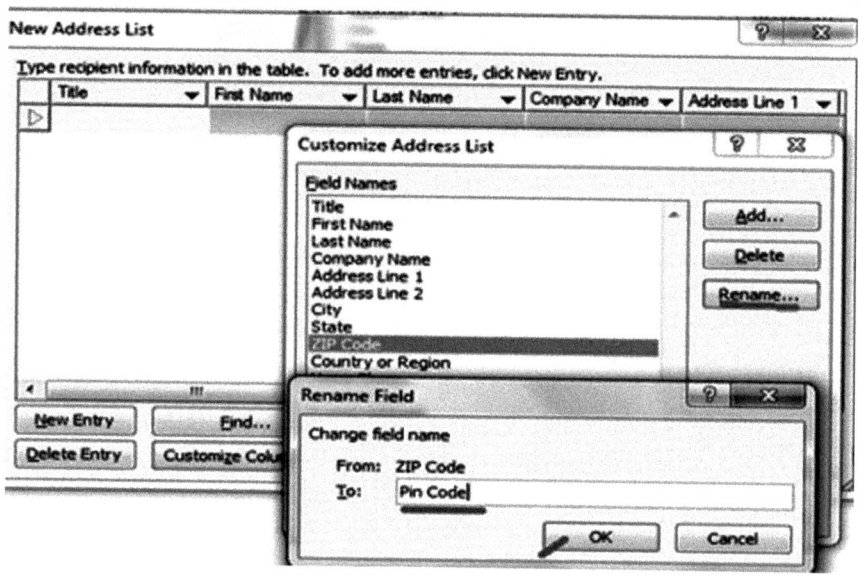

Clicking on OK button will change the name of the field. To remove the field select the field "E-mail Address," click on the delete button, and the field will be removed from the structure.

New Entry Button: This allows the user to enter a new record in the existing structure of records.

Delete Entry Button: This allows the user to remove a record that is already there in the file.

Find Button: This allows the user to search a particular record as per criteria defined by the user.

Step 2: Make the entries in the field structure as shown in the following figure:

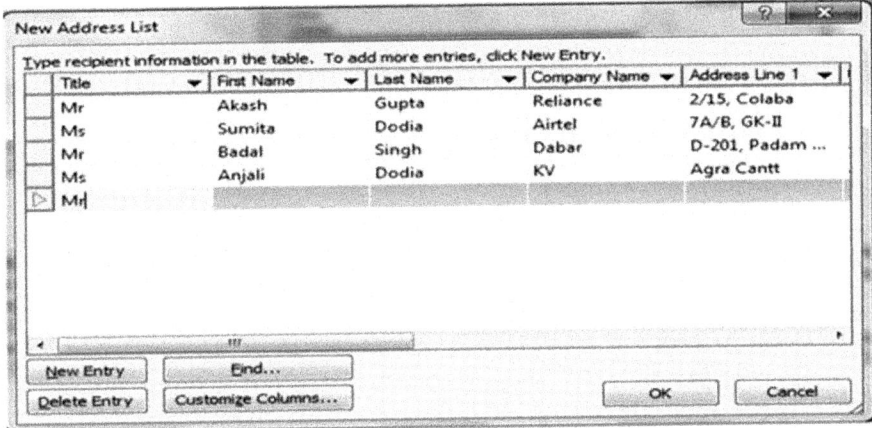

Once all entries are made, click on "OK." The mailing list is now ready, and clicking on OK will open up a new dialogue box (as shown in the next figure) where the user needs to specify the location where they want to save the address list and the name of the address list. Once this is done, the user should click on "save" to save the list.

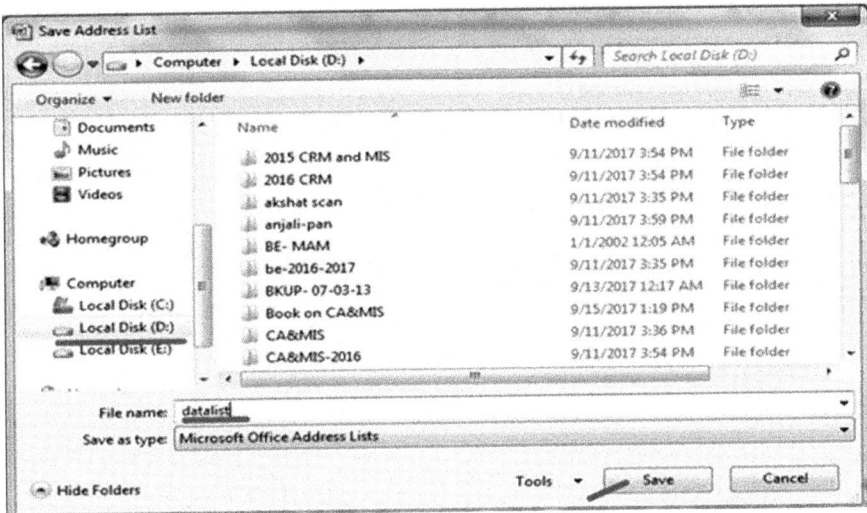

Step 3: Click on the "Start Mail Merge" button and select the document that you want to create from the drop-down menu. In this case we select "Letter."

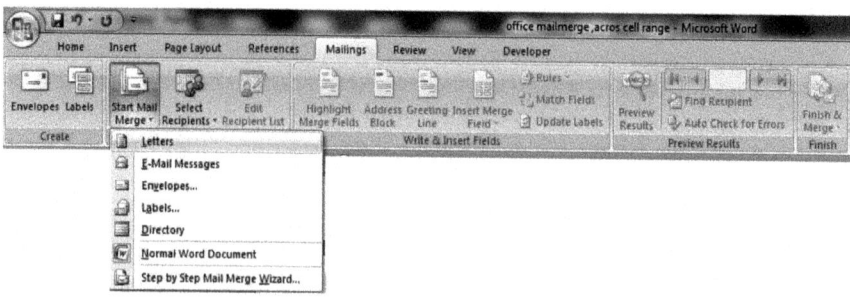

Step 4: A blank document will open and users should type the common text that is to be sent to all the members of the mailing list. Users draft a letter and to position the fields at the required position, users should first click on "select recipients," then click on "use existing list." A window will open asking the user to specify the location of the address list. They should define the location and click on "open."

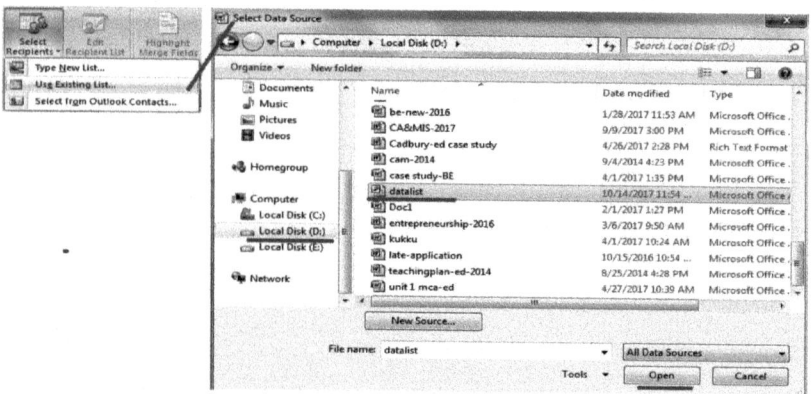

Step 5: After the execution of the Step 4, the user will find that few options on "mailing tab" ribbon become active, which helps in positioning the fields of the mailing list in the main document. The active buttons are shown in the following figure.

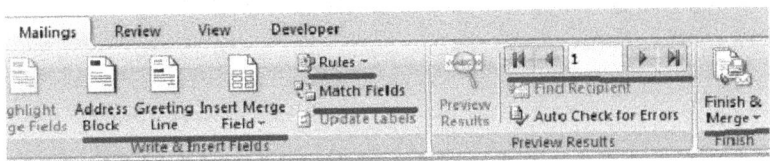

Address Block: Inserts all the fields in the main document from the mailing list.

Greeting Line: Inserts a greeting line in the main document.

Preview Result: Shows all the merged documents on screen.

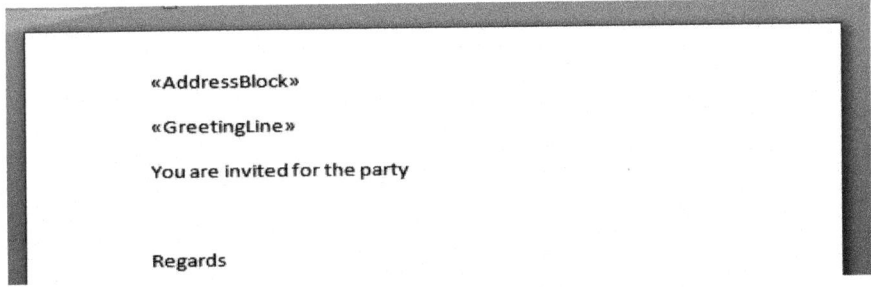 : Option lets the user see the preview of all the records one by one. The double arrow button shows the last record and the first record.

«AddressBlock»

«GreetingLine»

You are invited for the party

Regards

Clicking on Preview Result will generate the output as shown below:

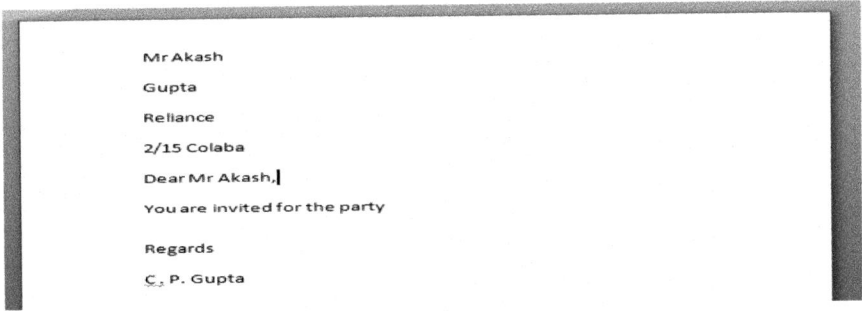

Mr Akash

Gupta

Reliance

2/15 Colaba

Dear Mr Akash,

You are invited for the party

Regards

C. P. Gupta

Insert Merge Field option allows the user to select the required fields from the structure of the mailing list and paste in the main document as shown in the following figure:

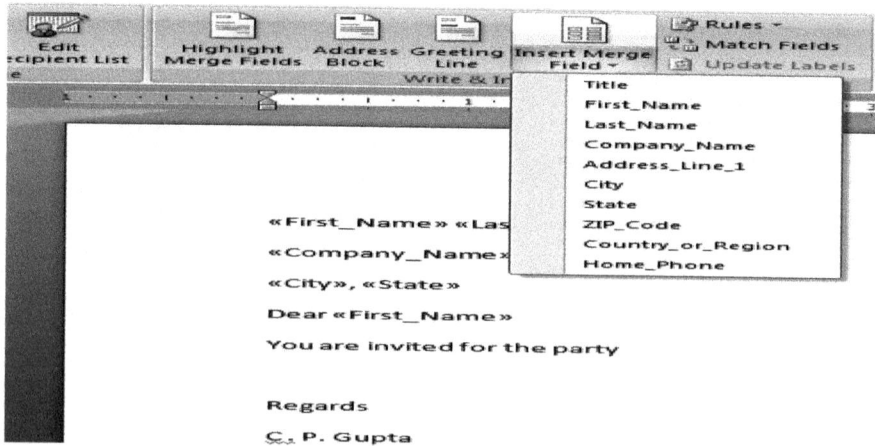

Clicking on "preview result" will generate the following result:

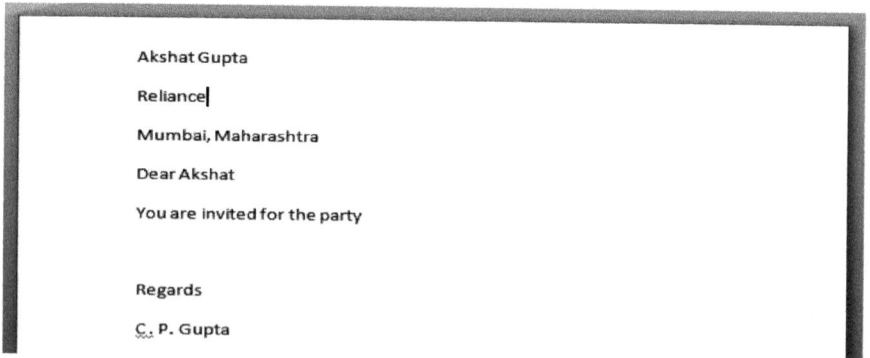

Finish and Merge option will generate a new file of all the merged letters. Number of letters displayed will be equal to the number of records in the mailing list.

Macros: While working in Word, users need to repeat certain steps. Instead of repeating them, users can automate them by creating and running macros. A *macro* is defined as a series of commands and instructions that a user can execute in one command to fulfil the task automatically. This bundling of commands help in saving time. There are certain steps that users need to take to record and run a macro. Word provides two options to run and record a macro; either by clicking on the "view" tab and then clicking on the "macros" button, or by clicking on the "window" button, clicking on the "Word option," and then selecting the checkbox in front of the "show developer: tab on the ribbon. This will make the developer tab appear on the ribbon. Here we learn a macro with the help of the "view" tab.

Step 1: Go to "view" tab, click on "macro," and "record macro." A dialogue box will open where the user needs to give the macro a name as in our example where we have given the name "first." Now to execute a macro either a button or a keyboard shortcut can be assigned.

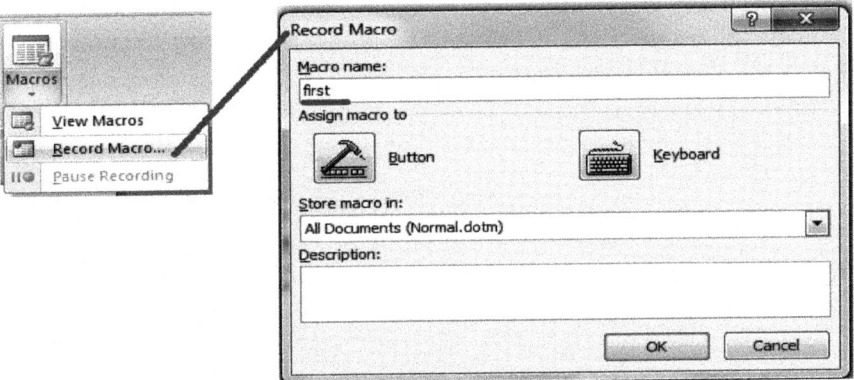

Step 2: Clicking on the "button," a screen will appear as shown in the following screenshot:

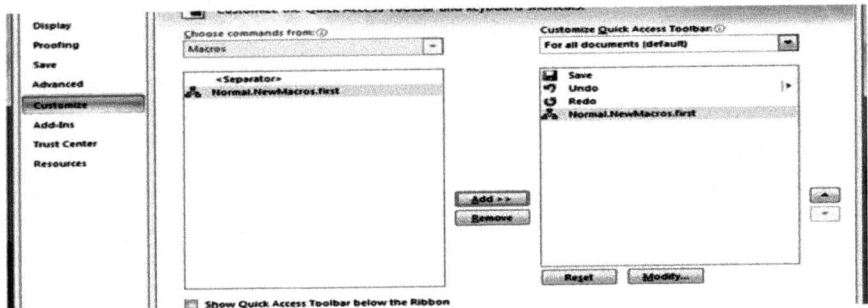

Select "Normal.NewMacros.first" and click on "add.: This will make the same phrase appear in the adjacent window. Now click on "Modify". A new window will pop up with number of icons as shown in the following figure. Selecting any one will assign a "button" to the macro created. Here we have selected "$" sign.

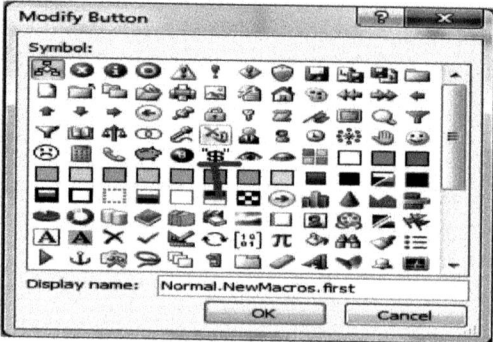

On the other hand a shortcut key can also be assigned to a macro in the following manner. In place of clicking on the "button," click on "keyboard and a screen will appear as shown here:

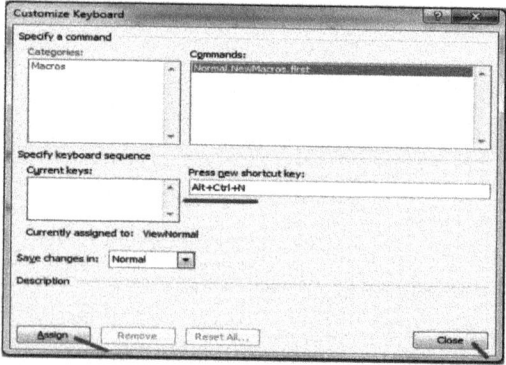

The macro will then record all the steps taken by the user. For example, here we have recorded on the first line "Hello," and on the second line "How are you." Click on the "Macros" button and click on the "stop recording" option. The view will be as shown in the next figure:

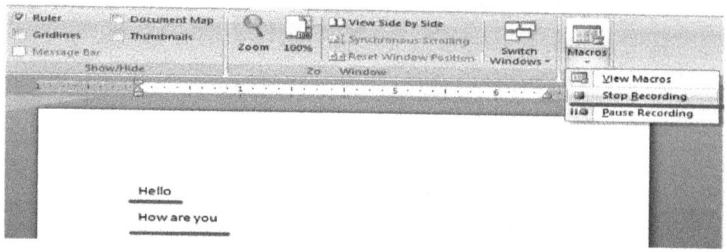

The macro is recorded and can be executed in two ways. First it appears as a button on the "quick toolbar." Clicking on it will execute the macro

and secondly by clicking on "Macros" and then on "View Macros." The following window will appear. Select the desired macro and click on "Run" to execute the macro. Another way of executing a macro is by creating a keyboard shortcut. Click on the

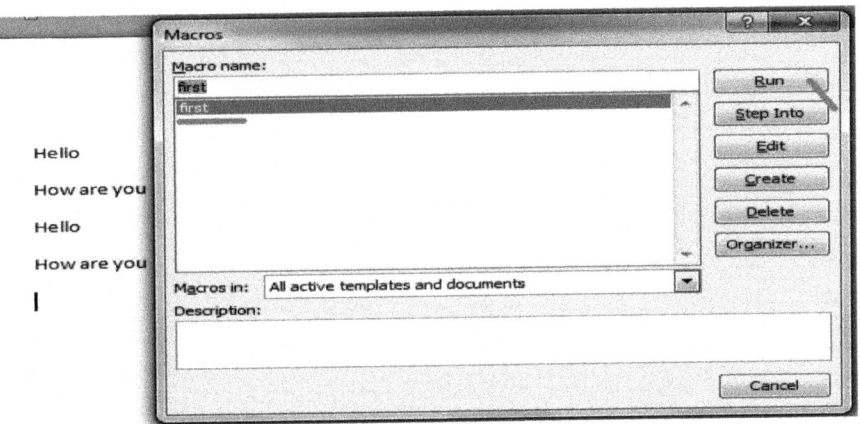

Thirdly, the macro can be executed by the shortcut key combination assigned. In our case it is Ctrl + Alt + N.

Cell Referencing and Range: Cell referencing means addressing a cell by its address. The cell in a table can be referred to as CnRn where C stands for column and R stands for row. The cell that contains the formula is not included in the calculation that uses a reference. If the cell is part of the reference, it is ignored. The first column of the table is referred to as A, the second as B, and so on. The first row of the table is referred to as 1, the second as 2, and so on. Thus the cell address of the fourth column and the third row will be D3. Range is specified as the number of cells. For example, we are adding cell A2 to D2 so the range is specified as (A2:D2)

Marks 1 (A1)	Marks 2 (B1)	Marks 3 (C1)	Marks 4 (D1)	Total
45	46	56	54	201
56	55	54	55	220

3.3 What is Microsoft Excel?

Excel is application software and is a part of the MS-Office package. Excel is used to create and edit spreadsheets that are in row and column format. The intersection of each row and column is called a cell and is used to hold either text or numerical values. Excel is used for tabular data calculation, creating pivot tables, and provides various graphing tools for data analysis. For example, a user can create a monthly budget, track sales figures, represent data graphically, or do database manipulation.

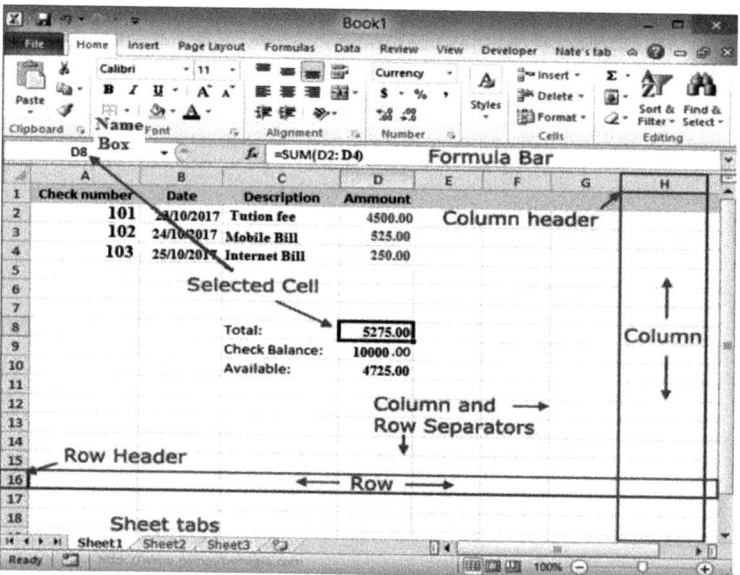

By default a workbook consists of three (3) sheets. There are 10,48,576 rows and 16,384 columns. A column can hold a maximum of 255 characters and a row height can be a maximum 409 points.

A key combination of Ctrl plus arrow keys is used to move across the sheet. For example

- Pressing **Ctrl** + ⇩ key take the cell pointer to the last row of the active sheet

- Pressing **Ctrl** + ⇧ key will take the cell pointer to the first row of the active sheet

- Pressing **Ctrl** + ⇦ key will take the cell pointer to the first column of the active sheet

- Pressing **Ctrl** + ⇨ key will take the cell pointer to the last column of the active sheet

- Pressing the home key brings the cell pointer to the first cell known as is A.

- Pressing the page down key will take the cell pointer to the row next to the last row of the present portion of the sheet displayed on the screen.

- Pressing the page up key will take the cell pointer to the first row of the previous portion of the sheet displayed on the screen.

Cell: The intersection of a row and a column is called a cell, and every sheet starts with cell A1. The active cell is the one whose border is darker than the other cells, and this is a place where the typed matter will appear in a sheet.

3.3.1 Cell Addressing

A cell address is specify in terms of CnRn, where C stands for column name and R stands for row number for example D5 means that column is "D" and the row is "5." In Excel three types of cell addressing is possible:

Relative Cell Addressing: Relative cell addressing means that the cell address will change as it is copied or moved, and the cell reference is relative to its location. For example, if in a cell D4 formula is written as =sum(A4..B4), when this is copied in cell D5 the formula will automatically change to =sum(A5..B5).

Absolute Cell Addressing: Absolute cell addressing means that the cell address will not be changed when the cell is copied or moved to another location. This is done by anchoring the row and column, and for anchoring the "$" symbol is used. For example, if formula D5 is used, copying this to another cell will not change the address.

Mixed Cell Addressing: In mixed cell addressing either the row or the column is made static i.e., fixed, so that when the cell is copied or moved any one value will change while the other will remain static. For example, copying D$5 to another location will change the column from D to the other, but the row will not change—it will remain "5." On the other hand if it is $D5, so copying it to other location will change the row number from 5 to any other but the column will remain "D".

3.3.2 Range

A range is a group or block of cells that are used

- to enter as an argument for a function;
- to define values to create a graph;
- to create *bookmarks* to specific data in a *workbook*.

Range can be defined as a group of cells that are selected for a purpose. The selected cells are surrounded by an outline or border as shown in the following figure. After selecting cells, right click on the mouse, a drop-down menu will appear, click on "name a range," a window will appear, enter the name of the range in the "name" column and press on OK.

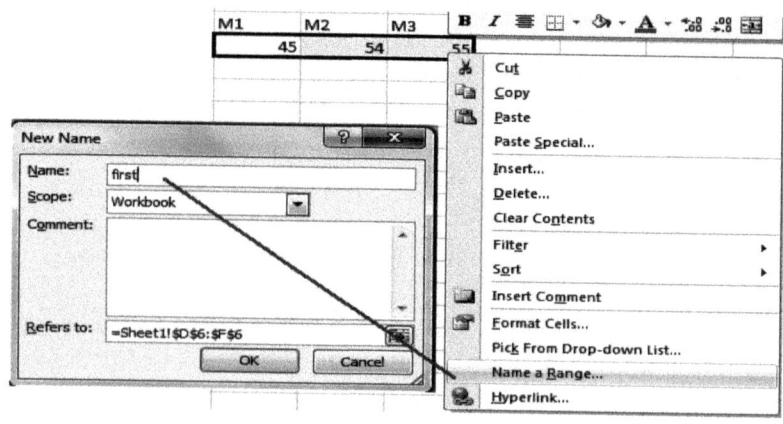

When we want to perform a function with the group of cells we can simply type the name of the range in place of their addresses as shown in the following figure. This simplifies the process.

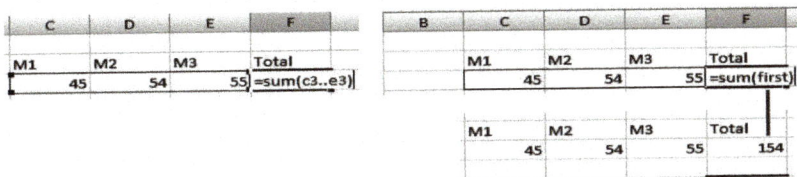

When a range consists of more than one cell, changes in the values of cell may affect all cells or some cells in the selected range.

Contiguous and Noncontiguous Ranges: When cells in the range are adjacent to one another it is called a contiguous range of cells. For example in the previous image cells selected in a range are C3, D3, and E3. This is an example of a contiguous series. When the selected cells in the range consist of two or more separate blocks of cells and are separated by rows or columns they are termed as a noncontiguous range. An example of this would be if one set of cell in the range is C4, D4, E4 and other is C6, D6, and E6.

Both contiguous and noncontiguous ranges can include hundreds or even thousands of cells and span worksheets and workbooks.

Selecting a Range: A user can select a range in a number of ways with the help of a mouse, or with the keyboard using the shift and arrow keys, or in the name box.

3.3.3 Changing the Row Height or Column Width

A column in Excel can take a width from 0 to 255. These numbers represent the number of characters that can fit in a cell. The default column width is of 8.43 characters, and if the column width is 0 the cell will not come into display, rather it will become a hidden cell. A row in Excel can take a height from 0 to 409 points. The default row height is 12.75 points and if the row height is 0, the row will disappear from view. There are number of ways by which row height or the column width can be changed.

First: Position the mouse pointer on the column header (in our example it is "E") or the row header, now right click on the mouse and from the drop-down menu select the "column width" or the "row height" option. Clicking on it will open a window asking the user to enter the new value for a column width or row height as shown in the following figure:

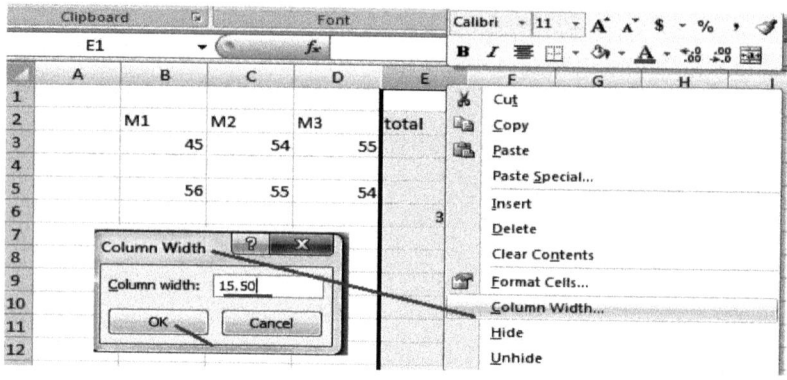

Clicking on OK will change the column width from a default 8.43 to 15.50.

Second: The same steps can be performed with the help of a mouse. To change the column width or the row height of any column or row, place the mouse pointer on the boundary of the cell and drag the boundary below the row heading or to the right of the column heading until the desired row height or column width is achieved.

To change the column width or row height of multiple columns or rows, the user needs to select all the columns or rows that they want to change, and then drag the boundary of any one of the selected column or row. The width or the height of all the selected columns and rows will change. On the other hand, if a user wants to change the column width or the row height for all the columns or the rows in the sheet, they need to click the "select all" button, and then drag the boundary of any on the column or the row. All column widths or row heights will change.

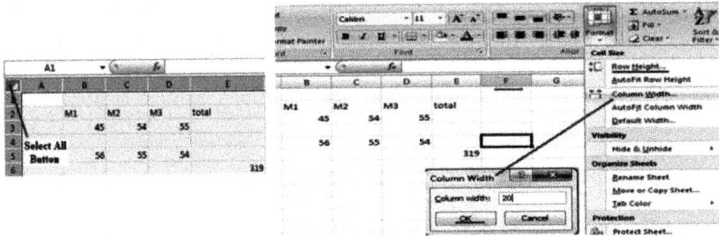

Third: Place the cell pointer in the cell (F in our example) of which the user want to change the row height or the column width. From the "home" tab click on the "format" button; a drop-down menu will appear as shown in the next figure. From this menu the user can select the "column width" option or the "row height" option to bring the desired changes in the sheet. If the number of rows and columns are selected, the change will occur in all of them.

3.3.4 Cell Alignment

Cell alignment refers to the positioning of the text or numbers in the cell. The user can align the text or numbers vertically: meaning toward the top, the middle, or the bottom of the cell; and horizontally: meaning to the left, the center, or to the right of the cell. This can be done in the following manner.

First: Select the cell/cells that consist of the text or numbers to be aligned. Go to the "home" tab, and choose from the "alignment" options as shown in the following figure. The text or numbers will be aligned accordingly.

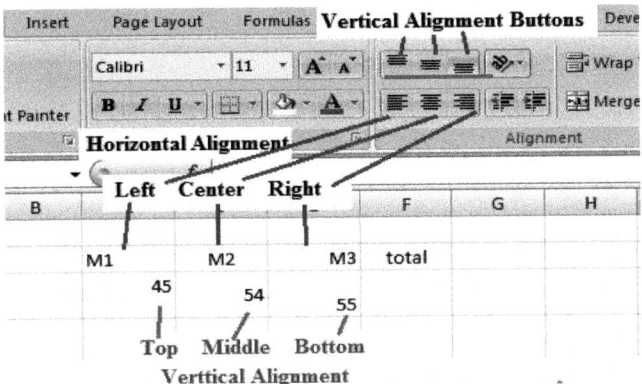

The same can also be done with the help of the alignment window as shown in the following figure:

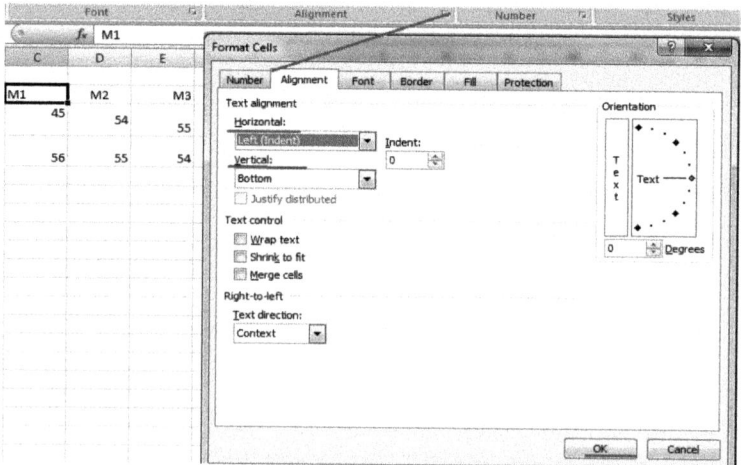

Wrap Text: When text typed in the cell does not fit, a part of the text might not be visible. To solve this issue without changing the width of the column, the user can place the mouse pointer on the cell and click on the "wrap text" button.

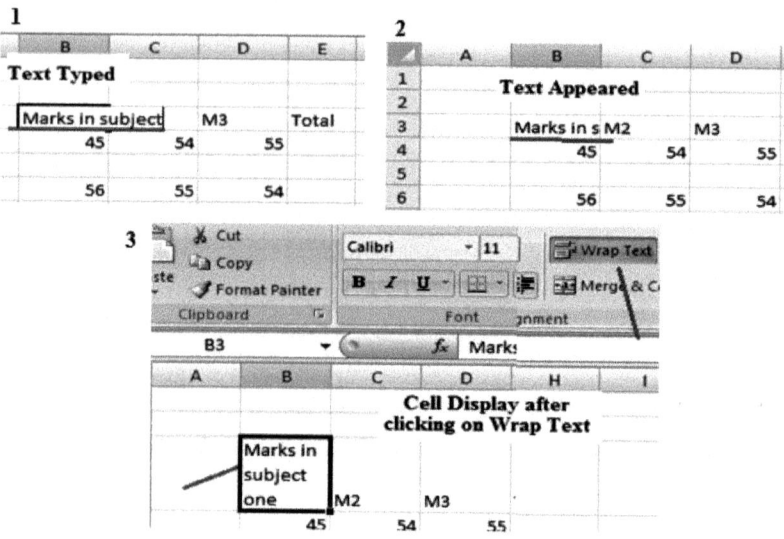

The same can also be done with the help of the alignment window by clicking on the check box in front of wrap text option.

Merge and Center: This option is used to center text that runs across several columns or rows. The user needs to select the cells and then click on "merge & center" as shown here:

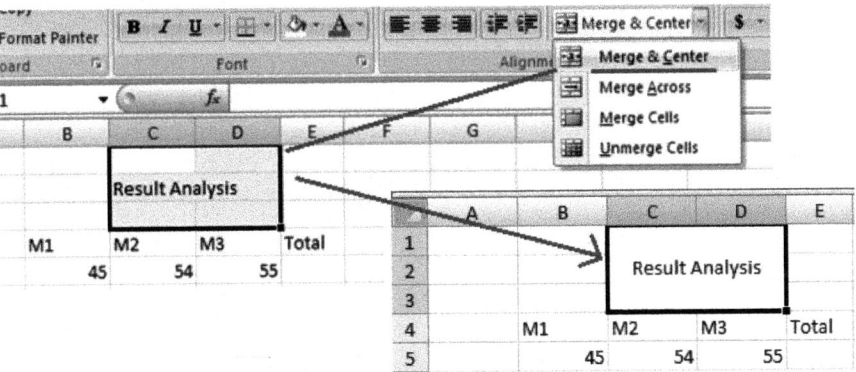

3.3.5 Sheets

By default, three sheets are there in a workbook. The number of sheets in a workbook depends on the free memory available. Certain operations can be performed on the sheets. When the user right clicks on a sheet the following menu will appear as shown in the following figure.

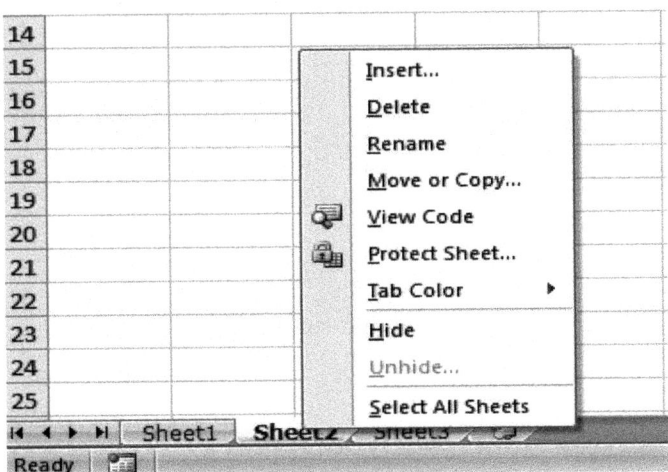

Insert: Clicking on insert will let the user insert a new sheet or certain objects in the active sheet.

Delete: Clicking on delete will delete the active sheet. In our example it is "Sheet 2."

Rename: will allow the user to give a new name to the selected sheet.

Move or Copy: will let the user to move the position of the sheet or create a duplicate copy of the sheet in the same workbook.

Tab Color: will change the background color of the active tab. In our case it is "Sheet 2".

Hide: will make the active sheet disappear from the display as show in the following figure:

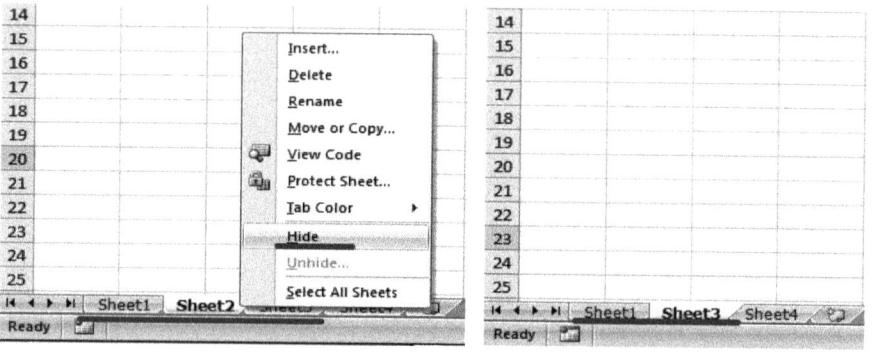

Unhide: will make the hidden sheet appear again as shown in the following figure:

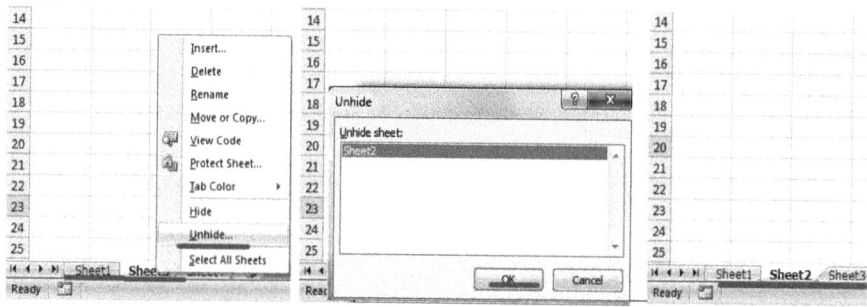

3.3.6 Creating and Inserting Charts in Excel

Charts are the graphical representation of the numeric data series. It helps in the easy understanding of the relationship between the different series of data. Excel provides users with a number of chart types to represent data. A combination of two or more chart types can also be produced. To create a chart, the user first needs to enter the data in the Excel sheet. First let us understand the types of chart provided by Excel.

Chart Types Available: Let us create a data series first to understand all the chart types.

		Quarter 1	Quarter 2	Quarter 3	Quarter 4
	Salesman 1	11	24	32	40
	Salesman 2	12	21	30	38
	Salesman 3	14	23	28	43
	Salesman 4	12	25	35	45

Here is a sheet that consists of data for four salesman for all the four quarters. The user can select all four data series or they can select nonadjacent rows in the data series by pressing Ctrl Key and drag the mouse to select the data series. All charts provided by Excel can plot data on single or multiple data series. However, pie charts work with a single data series.

Pie Chart: Pie chart represents only one data series. Even if we select the full table, the pie chart will show the graphical representation of the first series only. The subtypes of a pie chart are as shown here and are represented by 2D, 3D, and exploded format.

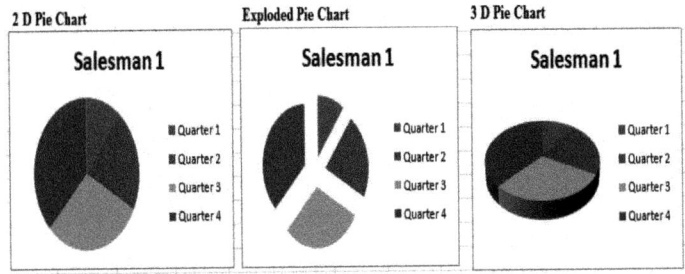

There are also two other available formats called "pie of pie" and "bar of pie."

Column Chart: A column chart is the most preferred chart to present data. It is used for making a comparison when there is more than one data series. The horizontal line is termed as X axis and the vertical line is termed as Y axis. In a column chart, the vertical axis (Y-axis) is used to display numeric values, and the horizontal axis (X-axis) is used to display the other category. For example, the following chart shows the comparison of the sales performance of all four salesmen for each quarter:

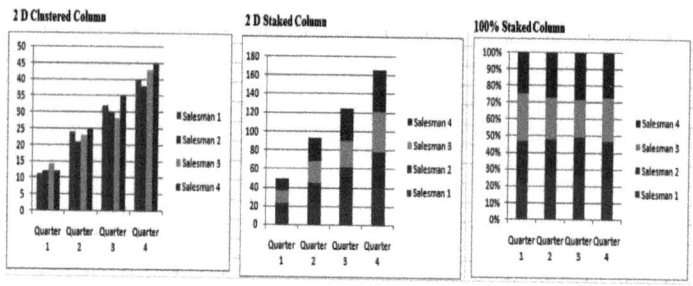

The column chart has also got its subtype. The above image shows a 2D clustered column chart, 2D, staked column, and a 2-D 100% staked column charts.

A Stacked Bar Chart, also known as a stacked bar graph, is a graph that is used to break down and compare parts of a whole. Each bar in the chart represents a whole, and segments in the bar represent different parts or categories of that whole as shown in the following chart:

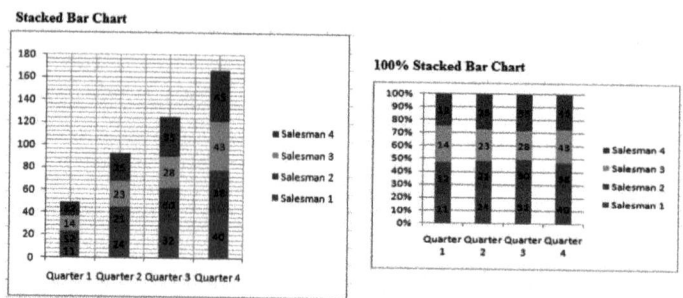

On the other hand, a 100% stacked bar chart is an extension of the stacked column chart in that it compares the percentage that each value contributes to a total as shown in the preceding chart.

The next category is the 3-D column chart; in it subcategories are clustered and stacked, 100% stacked and 3-D Column. The available options are cylinder, cone, and pyramid. The subcategories are clustered, stacked, 100% stacked, and 3-D.

Line Chart: The line chart is used to display trends. The horizontal line is termed as X-axis and the vertical line is termed as Y-axis. In a line chart, the vertical axis (Y-axis) is used to display numeric values, and the horizontal axis (X-axis) is used to display other category.

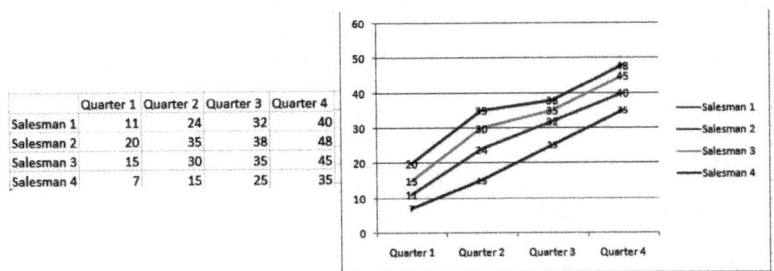

	Quarter 1	Quarter 2	Quarter 3	Quarter 4
Salesman 1	11	24	32	40
Salesman 2	20	35	38	48
Salesman 3	15	30	35	45
Salesman 4	7	15	25	35

Other categories include stacked line chart, 100% stacked line chart with or without markers. A 3-D line chart can also be made, however it does not display data well in three dimensions.

Bar Chart: A bar chart and a column chart are the same thing. The only difference is the way the data is presented in the chart. In a bar chart data is presented using horizontal bars, and in a column chart data is presented using vertical bars. The horizontal axis of a bar chart contains the numeric values. The length of each bar or column is proportional to the data that it represents, so a bar or column corresponding to a value of 100 would be twice as long as one corresponding to a value of 50.

Presenting the Same Data Series

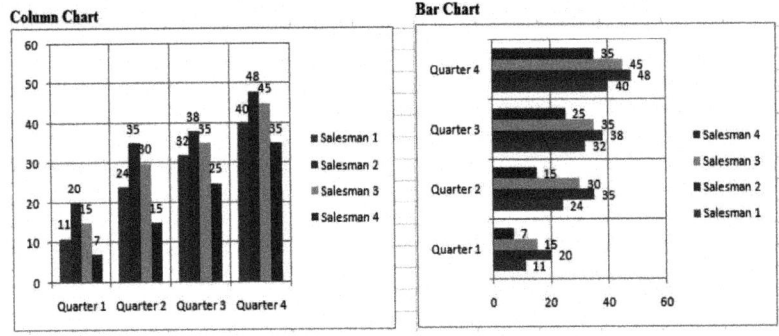

The next category is of 3-D bar chart; in it subcategories are clustered, stacked, 100% stacked. Then available options are in cylinder, cone, and pyramid with subcategories of clustered, stacked, and 100% stacked.

Area Chart: Area charts and line charts are similar with the only difference being that the area below the plot line is solid. There are three charts available: the area chart, the stacked area chart, and the 100% stacked area chart. Each of these charts come in 2-D format and in true 3-D format with X, Y, and Z axes.

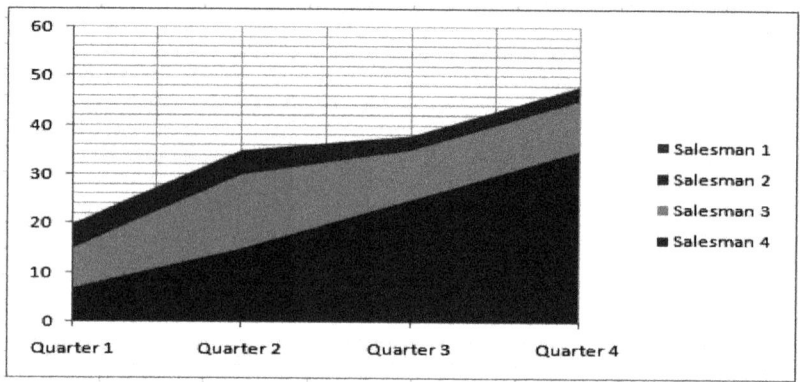

In case of multiple data series, the area charts are ineffective as they hide the series with lesser values behind the series with greater values. To remove this we can use a 2-D stack area chart, a 2-D 100% stack area chart, or a 3-D area chart as shown in the following chart:

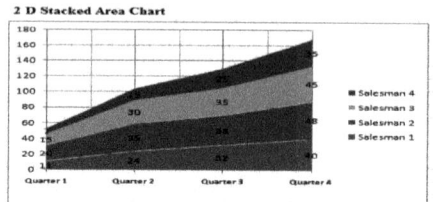
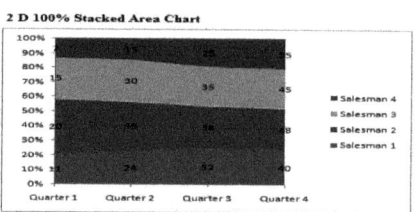

Scatter Chart: A scatter chart is the only chart where both axes represent numerical data. The main difference between scatter and line charts is the way that they plot data on the horizontal axis. Often referred to as an XY chart, a scatter chart always has two value axes to show one set of numerical data along a horizontal (value) axis and another set of numerical values along a vertical (value) axis. The chart displays points at the intersection of an x and y numerical value, combining these values into single data points. These data points may be distributed evenly or unevenly across the horizontal axis, depending on the data. This chart is used when the user wants to compare two sets of values for each series. For example, comparison of the height and weight ratio of a group of boys as show in the following figure:

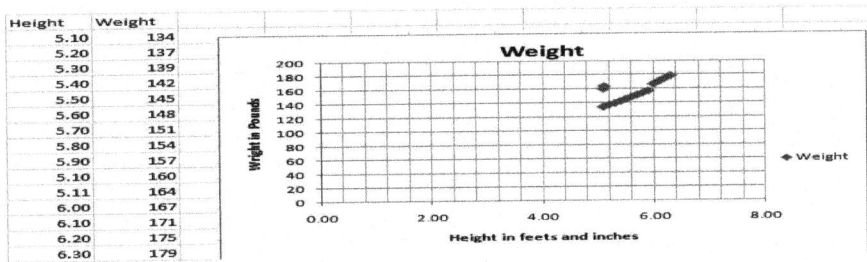

Height	Weight
5.10	134
5.20	137
5.30	139
5.40	142
5.50	145
5.60	148
5.70	151
5.80	154
5.90	157
5.10	160
5.11	164
6.00	167
6.10	171
6.20	175
6.30	179

The first data point to appear in the scatter chart represents both a y value of 134 (weight) and an x value of 5.10 feet's (height). Scatter charts are commonly used for displaying and comparing numeric values, such as scientific, statistical, and engineering data.

Line charts on the other hand can display continuous data over time, set against a common scale, and are therefore ideal for showing trends in data at equal intervals or over time. In a line chart, category data is distributed evenly along the horizontal axis, and all value data is distributed evenly along the vertical axis. As a general rule, a line chart should be used if data has non numeric x values. For numeric x values, it is better to use a scatter chart.

Other Chart Types: Apart from those types above Excel also presents Stock, Surface, Doughnut, Bubble, and Radar Chart Type.

Chart Elements: A chart has many elements. One can change the display of the elements of the chart by moving them from one location to another, resizing or by changing their format. They can also be removed. Various numbers represent the various areas of a chart as shown here:

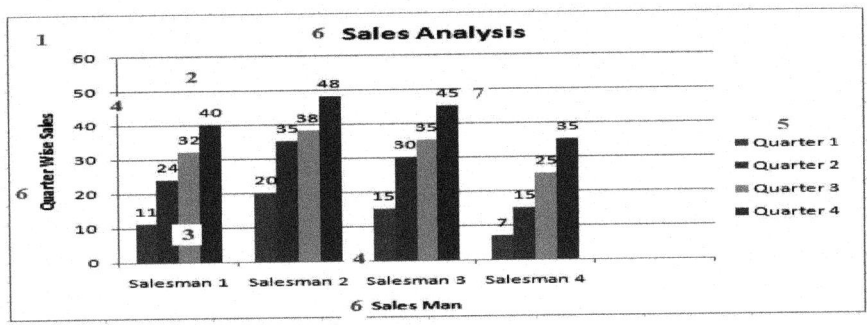

- the chart area of the chart
- the plot area of the chart
- the data series that are plotted in the chart
- the horizontal axis and the vertical axis
- the legend of the chart
- a chart and the axis title
- a data label that you can use to identify the details of a data point in a data series

How to create a chart: Enter the data series in Excel sheet. Select the data series, click on "insert," click on chart type, and select the required chart type from the drop-down menu. A chart will appear in the sheet as shown here:

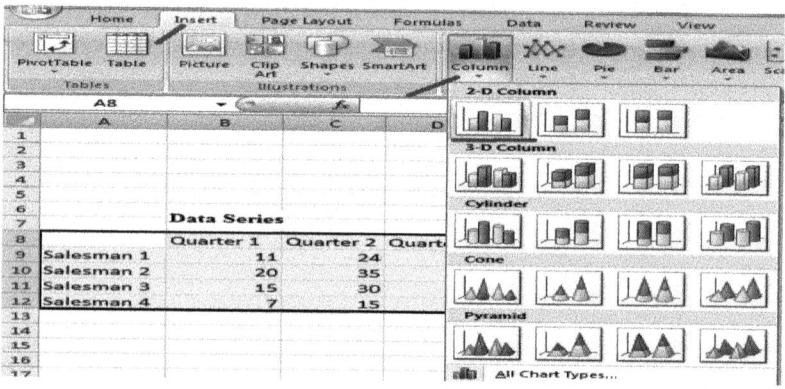

A chart will appear in the sheet as shown here:

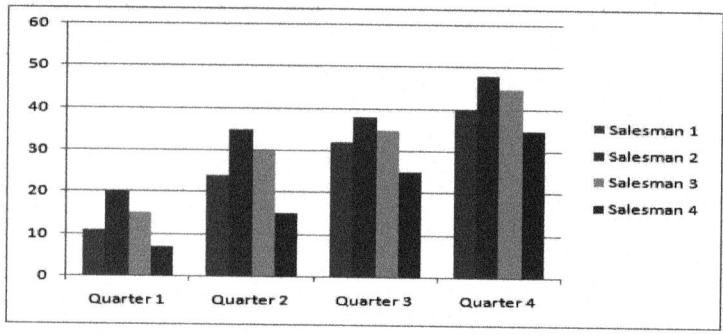

How to modify the chart: Once the chart has been created, the user can modify its elements. The following steps are to be taken to modify the chart.

Step 1: Click on the chart area to select it. After selecting the chart a "layout" tab will appear on the ribbon and will provide all the modification options as shown in the following figure:

- **Add Chart Title and Axis Title:** This helps in identifying the information that appears in chart. The axis title helps in the identification of the data series plotted on x and y axis as shown in the next figure below:

Step 1: Click on the "chart title" button; a drop-down menu will open. Select the option you want and a box will appear in the chart. Name the chart and it will appear in the chart as show here:

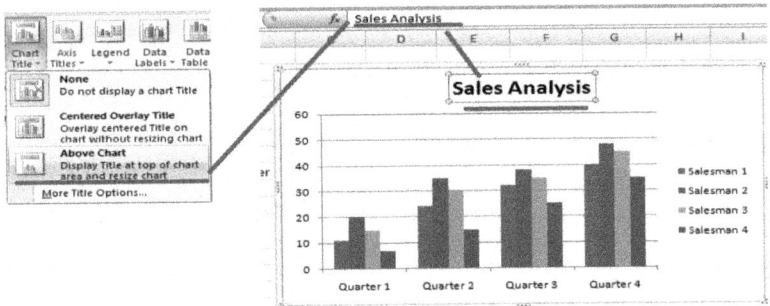

Step 2: Click on the "axis title" button, a drop-down menu will appear with two options: horizontal axis and vertical axis. Select the required option and type the name of the axis as shown here:

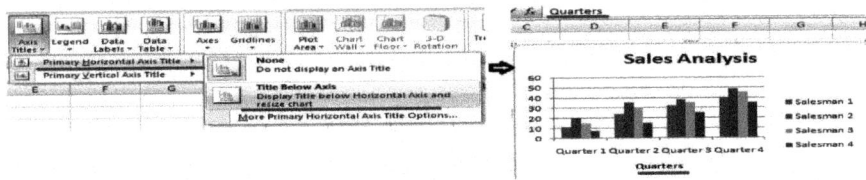

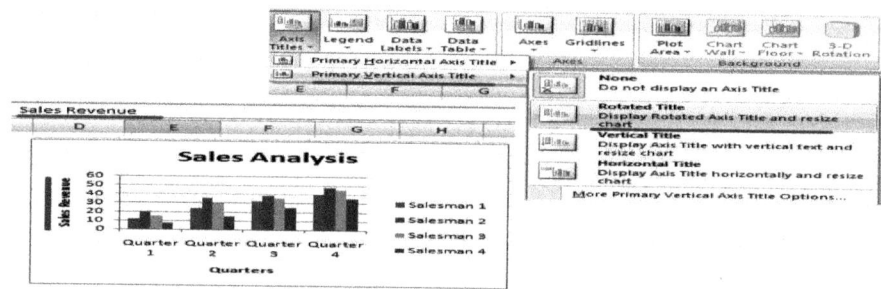

- **Add a Legend:** A legend represents the data series that are graphically represented in a chart. It appears by default on the right-hand side of the chart, but it can be modified and can be moved to other locations of the chart as shown here:

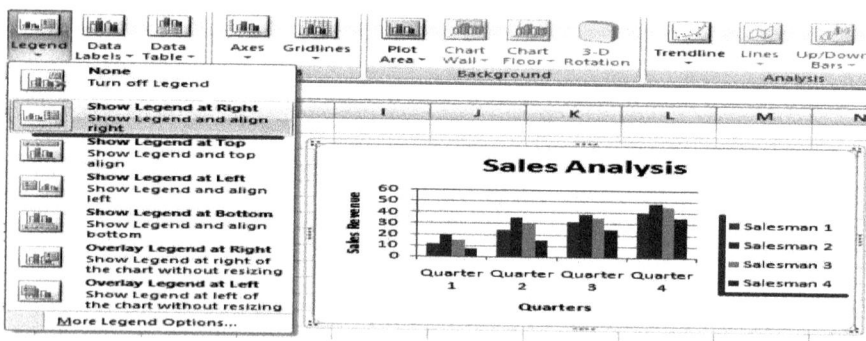

- **Add Data Labels:** Data labels are the values that are represented graphically in a chart. Adding data labels helps the user in identifying the values of the pictorial representation of the data series in a graph as shown here:

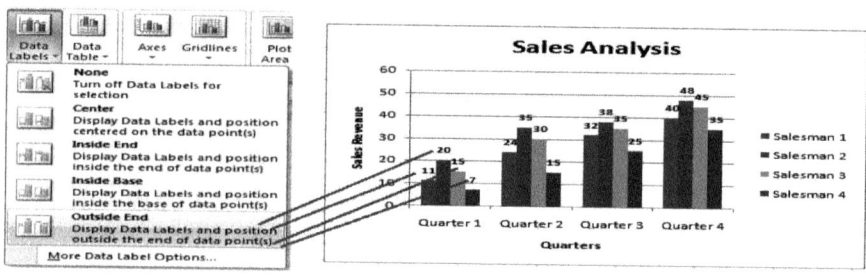

Add Data Table: Data table consists of all the values of the data series and can be added with the chart as shown in the figure below:

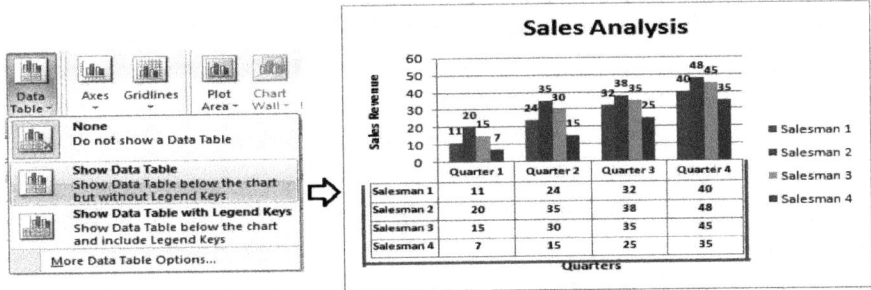

Add Formatting to a Chart: Every element of a chart can be given a different style with the following buttons available on the "format" Tab.

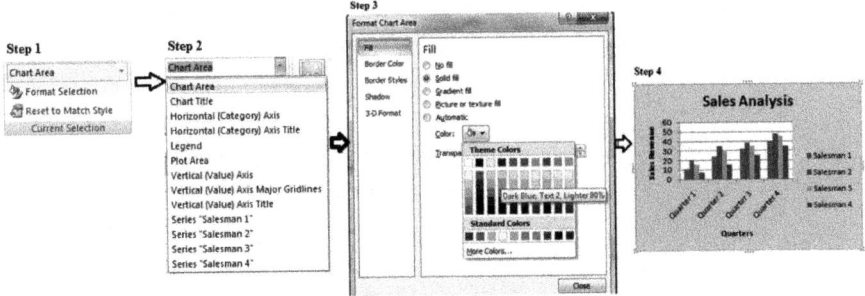

All the elements of chart that are visible in the step 2 diagram can be modified as per the user's need.

Design Tab for Chart: Design tab for chart gives the user the following option as shown in the next figure:

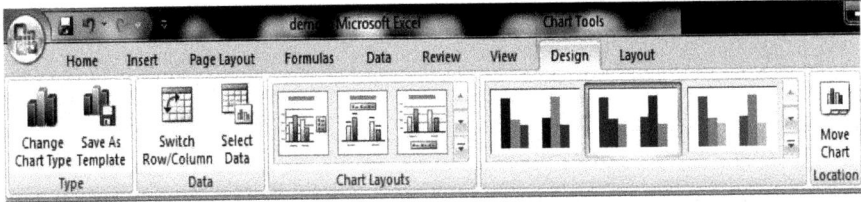

Change Chart Type: For activating the "design" tab select the chart first. Now click on the design tab and under it, click on "change chart type"

button. A screen will appear as shown in the next figure. From here, the user can click on the desired chart type and the chart will be modified accordingly.

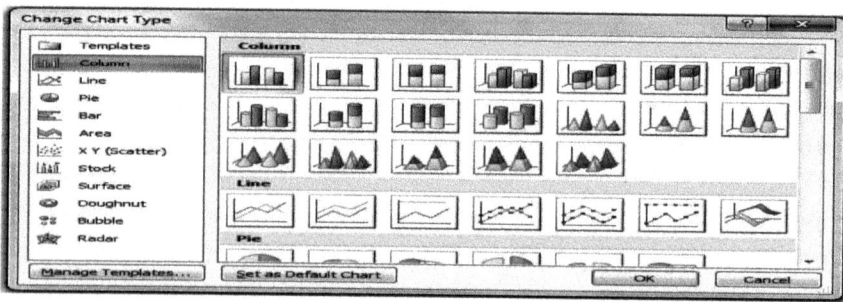

Save as Template: If the user wants to reuse a chart that he or she has customized, it can be saved as a chart template file and can be used to apply on the new chart that the user will create later. Select the chart, click on the "save as template" button and a window will appear asking the user to save the file as "chart template file" at the location specified by the user.

Switch Row and Column: This option will change the axis, i.e., X axis values will be shown on Y axis, and the Y axis values will be shown on X axis as shown here:

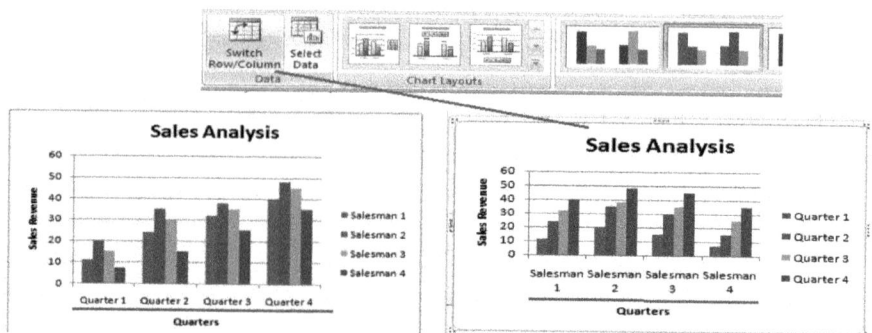

Select Data: This will allow the user to select the data source from which the chart will be made. The following steps are to be taken. Click on "select data" button and a window will appear in which the user can specify the data table range. Click on OK will change the appearance of the chart.

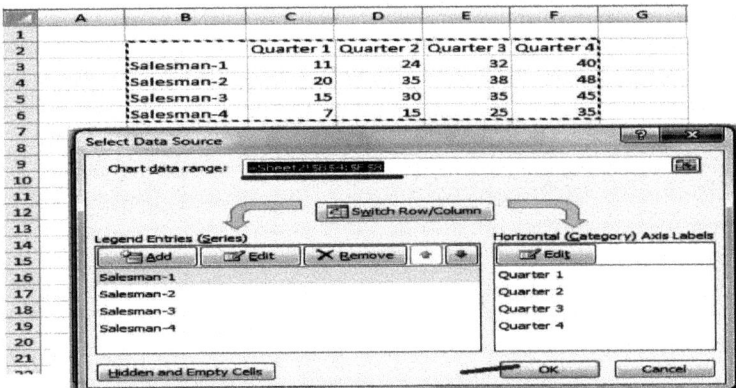

Clicking on the any chart style will bring the same change in the chart as shown here:

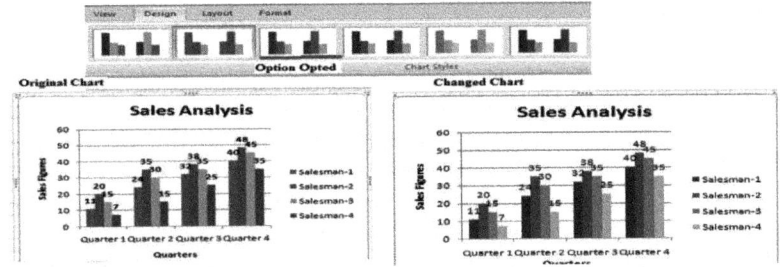

Move Chart: Helps to choose where the user wants to place their chart. It could be on the same sheet as an object or on the new sheet. To do this the user should select the chart, then click on the "move chart" button. A window will appear asking the user the position where they want to place the chart. The user selects the location and clicks on OK. The chart will be saved as a separate sheet or in the same sheet as an object as shown in the following:

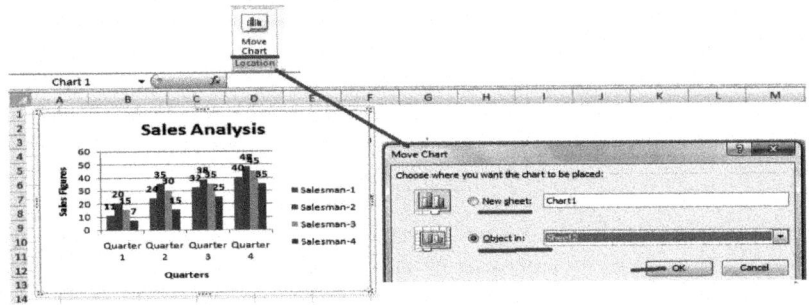

3.3.7 Sort and Filter

Sorting means arranging the text, numbers, dates, and time in a particular order. Text is arranged alphabetically either from A-Z (ascending order) or from Z-A (descending order). Numbers are arranged from lowest to highest (ascending order) or from highest to lowest (descending order). Date and time are arranged from oldest to newest (ascending order) or from newest to oldest (descending order).

Step: To sort the text the user needs to select a column with alphanumeric data or the active cell should be in the column where the text data is written or select the entire table (first row is taken as a header row and does not become the part of sorting). Click on the "home" tab and under it click on the "sort & filter" button. A drop-down menu appears asking the user to select the sort criteria as shown in the next figure:

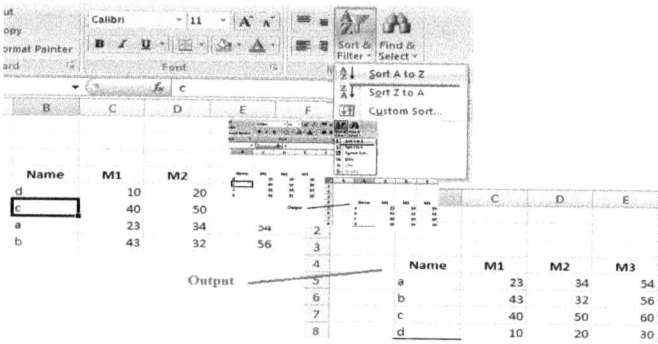

To sort the numbers repeat the same steps. The only difference is that now the active cell should be on any row of the column that consists of a number as shown in the next figure. Here the data is arranged in a descending order. Also the sort option changes from A-Z to smallest–largest and vice versa.

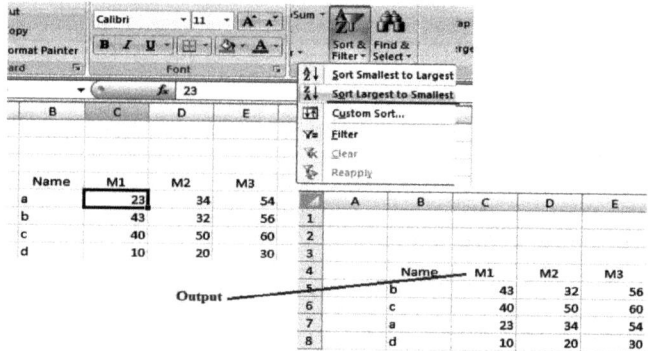

Custom Sort

This gives an option to sort the table as per the user's choice irrespective of the active cell pointer position. To execute this option the user first needs to select the table and then click on "sort & filter" and under it click on "custom sort." The screen will appear as shown in the following figure.

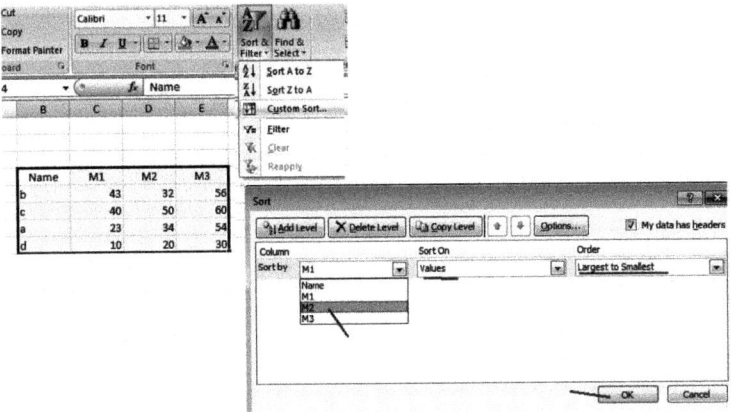

Clicking on "OK" will sort the table on values under head "M2" in a descending order.

Filter

The Filter command helps the user to list out only those records from the data sheet that match the specified criteria. For example, we want to see only those records for which a value of M2 is greater than or equal to 34. To do this the user first needs to select the table and under the "home" tab click on "sort & filter." Under that the user should click on "filter." on the user then clicks on the down arrow of the selected column to set the filter as shown in the following figure.

Step 1:

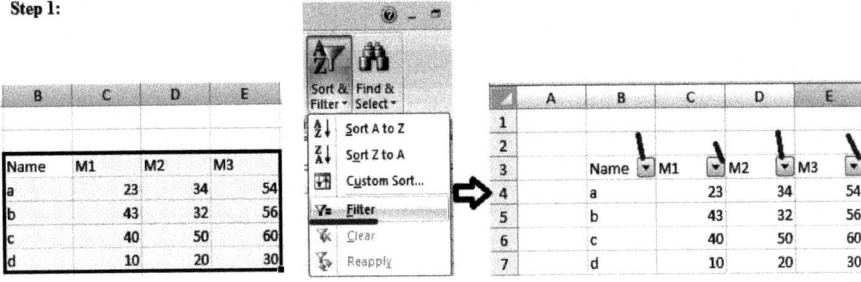

Click on the down arrow sign on M2, and a drop-down menu will appear. From this click on "number filter," and click on the option "greater than or equal to." Another window will open in which the user will write the value 34, as shown in the following figure:

Step 2:

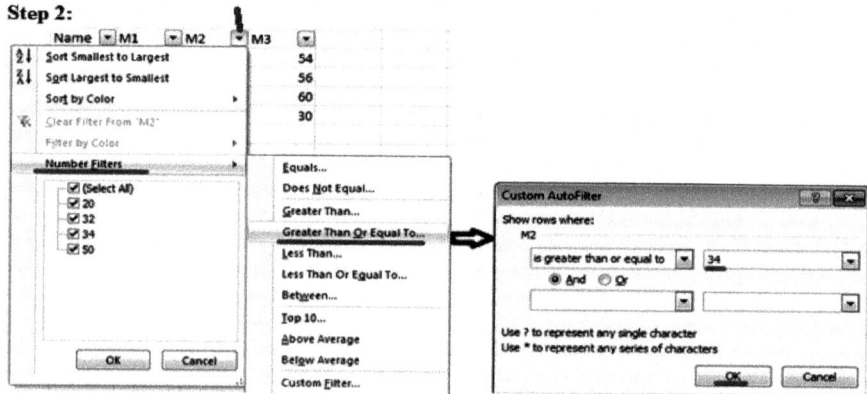

Clicking on OK will generate the required result as shown in the following figure:

Step 3:

	A	B	C	D	E
		Name ▾	M1 ▾	M2 ⊠	M3 ▾
1					
2					
3		Name	M1	M2	M3
4		a	23	34	54
6		c	40	50	60
8					

K18   *fx*

Setting Multiple Filters

Now we want to show the result for which M2 is greater than or equal to 34 and M3 is greater than or equal to 55. We shall repeat steps 1 and 2 of the filter process. The result of the previous process is step 3, and we shall be clicking on the down arrow at M3 as shown in the following figure:

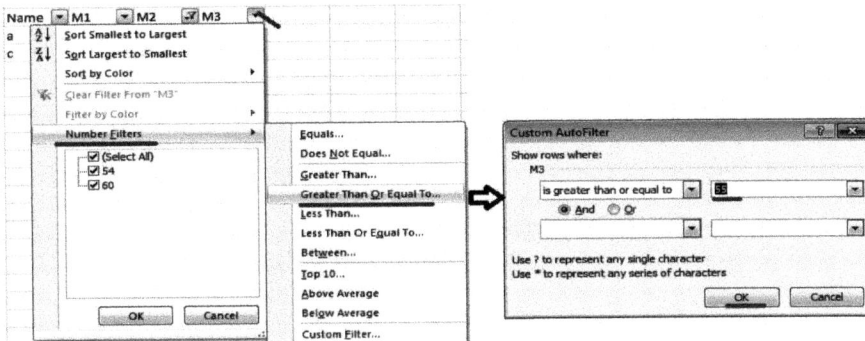

Clicking on OK will generate the result as shown in the following figure:

	A	B	C	D	E
1					
2					
3		Name	M1	M2	M3
6		c		40	50	60
8					

3.4   Working with Powerpoint

Microsoft PowerPoint is used as a software for presentation. It offers a user a number of formats in which slides called "slide layout" can be prepared. It provides the user a facility to insert graphics, movie clips, and sound into a slide and provides powerful tools for presentations.

3.4.1 Home Tab

Slide Layout: Slide layout can be defined as the positioning of different elements on the slide. By default the slide that opens first consists of a title block at the top of the slide and a text block below it. When the user inserts a new slide, a drop-down menu appears as shown in the following figure. From this menu the user can select the desired slide layout. If the user wants to change the layout of the slide, they can click on the layout button and then click on the desired option and the layout will change accordingly.

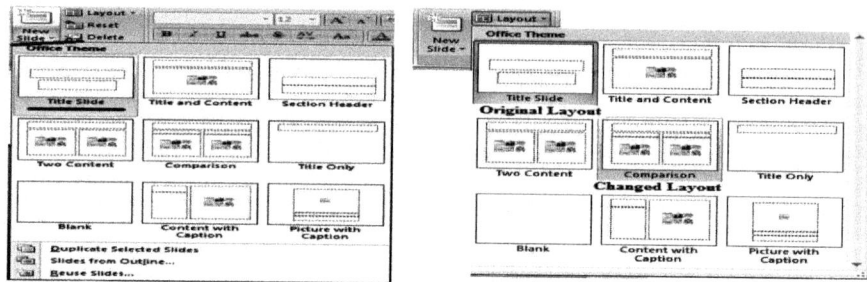

The delete button will delete the current slide. The font and paragraph options work the same as in Word.

Drawing: This lets the user insert the shapes available in the slide at a desired position. Steps to be taken are: click on any desired drawing image and then click on the slide. Image will appear in the slide at the mouse pointer position.

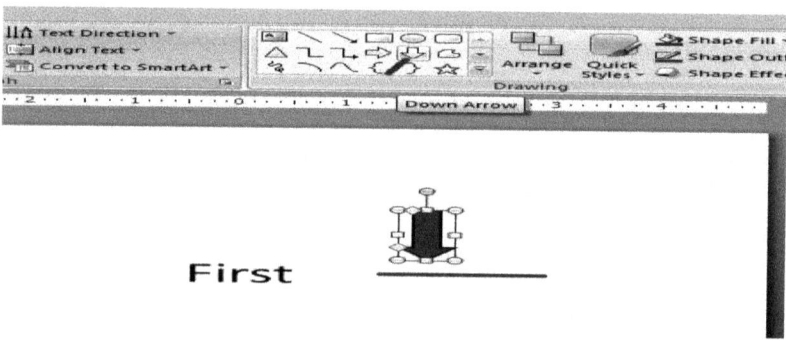

Arrange option: Select the inserted image. Place it at the desired position. Now click on arrange and click on the desired option, which in our case is "send to back." You can see in the next diagram that the image is sent back and all characters are now visible.

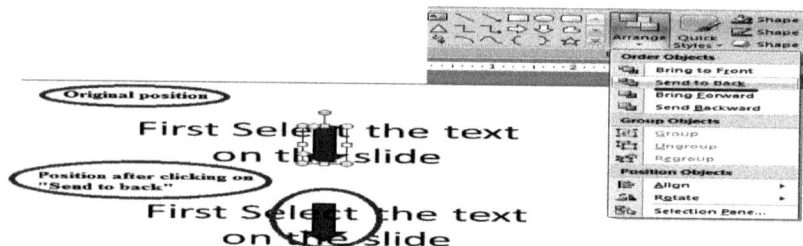

Quick Style: This helps the user to apply the previously available styles on the selected portion of the slide as shown in the next figure. First select the part of the slide where the style needs to be applied. Click on the "quick style" button and select the desired style. Change will appear in the slide as shown in the next figure:

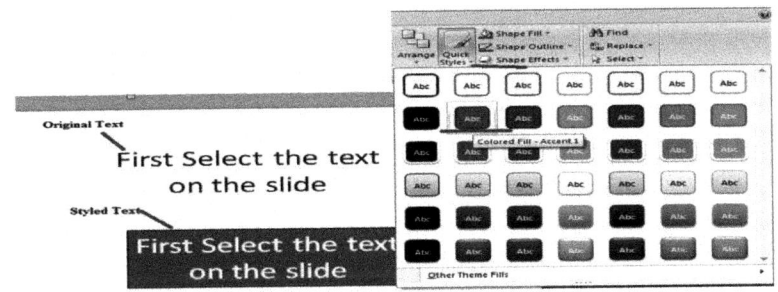

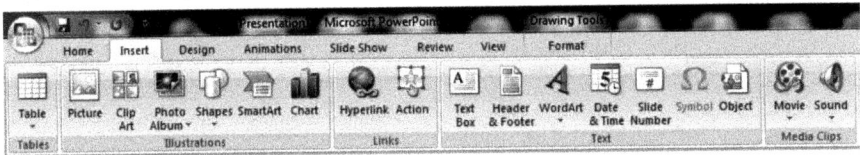

3.4.2 Insert Tab

Insert tab options are the same as they are in Word. A user can insert a table, picture, clip art, photo album, shapes, smart art, chart, hyperlink, text box, headers and footers, word art symbol, object, movie clip, and sound as learned in MS Word.

Date & Time: This allows the user to insert the date, time and page number in all the slides. Clicking on Update will automatically display the current date and time whenever the slide is opened, regardless of the time and date of its creation. Steps to be taken are: click on the insert tab, then click on the "date & time" button, click on the check box before "date & time," select the radio button "update automatically," and click on the check box "slide number." The changes can be applied to all the slides by clicking on the button "apply to all," or to the current slide by clicking on the "apply" button.

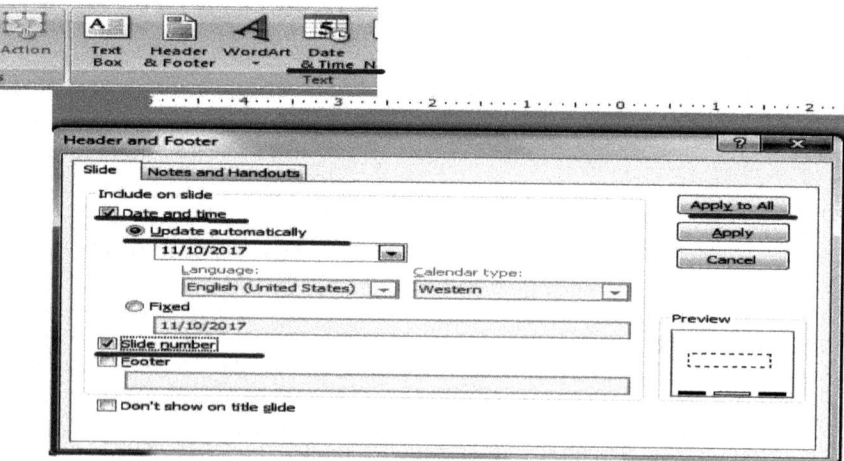

3.4.3 Design Tab

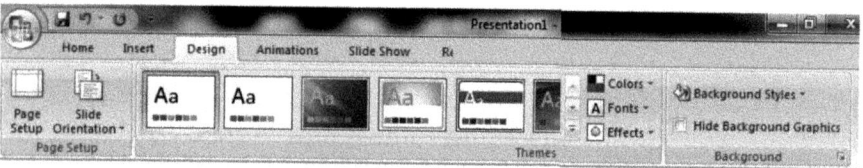

Page Setup: Allows you change the size of the slide. Click on the "page setup" button. A window will open asking the user to change the dimensions of the slide as shown in the following figure. Dimensions can be changed either by changing the width and height, or by clicking on the "slide sized for" option and then clicking on "OK."

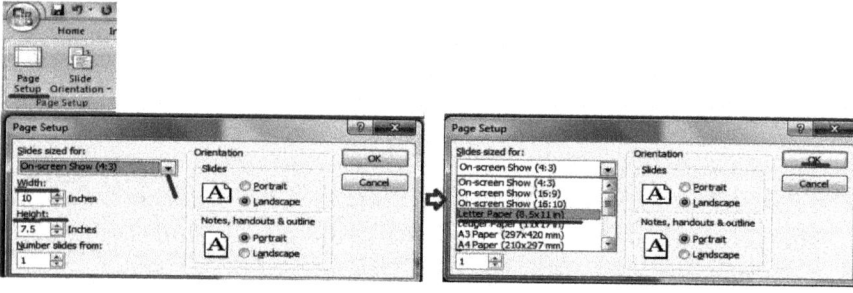

Slide orientation can be either portrait or landscape, just as in MS Word.

Themes: These are also called "templates" and are predefined designs that can be applied to the presentation.

3.4.4 Animation Tab

Slide Transition: When an animation-like effect is added to a slide it is called "slide transition." This means slides are appearing in an animated way; one is coming down from the top, another from left to right, and so on. Users can control the speed, add sound, and customize the properties of transition effects. Following are the steps to be taken:

1. Either place the mouse pointer on the slide or select the slide from the left-hand display of the window in slide thumbnails. Now in the "animation tab" select any of the options available on the "transition to this slide" options as shown in the following image. The selected effect will come on the slide, clicking on the down arrow as shown in the figure will give the user a number of options to select from.

More Options

2. Sound can be added to the slide transition by clicking on the down arrow on the "transition sound" option, and the slide transition speed can be controlled by clicking on the "transition speed" option and under it by selecting the desired sound or speed choice.

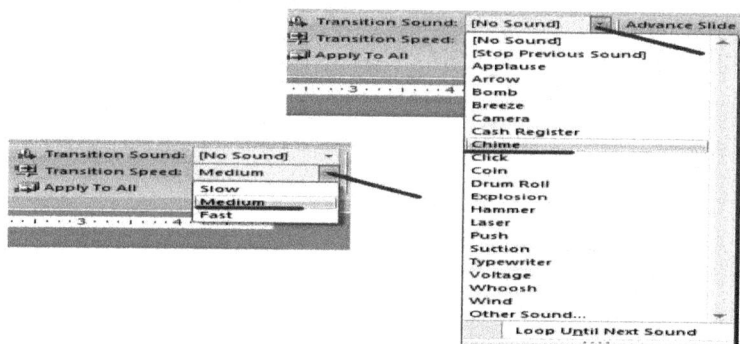

The movement from one slide to the next can be either by using the mouse or can be automatic by setting the time as shown in the following figure:

 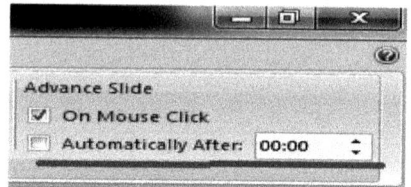

Remove a Transition Effect: Either place the mouse pointer on the slide or select the slide from the left hand display of the window in slide thumbnails. Now in the "animation tab" go onto the "transition to this side" option and in it click on the down arrow. A screen will appear with number of options under it. Click on "no transition," and the transition effect will be removed from the slide.

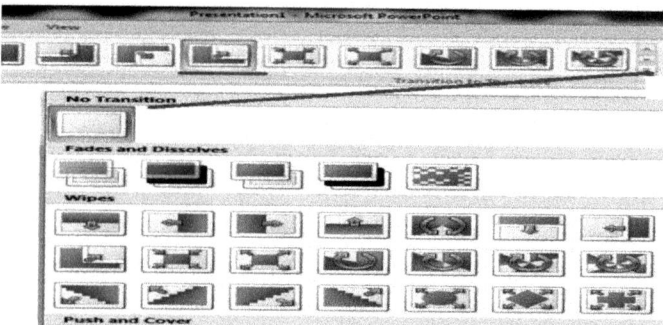

Text Animation: "Text animation" can be defined as an adding of a number of effects to the text. For example, moving letters, words, or paragraphs across the computer screen. There are a wide range of special effects that can be used with text animation, many of which are identical to traditional three-dimensional animation effects and two-dimensional effects. To apply the text animation, select the text that may contain a word, a sentence, a paragraph, or a complete slide matter. Now click on the custom animation option on the "animation tab". Two options will appear: one is "animate" and the second is "custom animation." The "animate" button gives some predefined text animation effects as shown here:

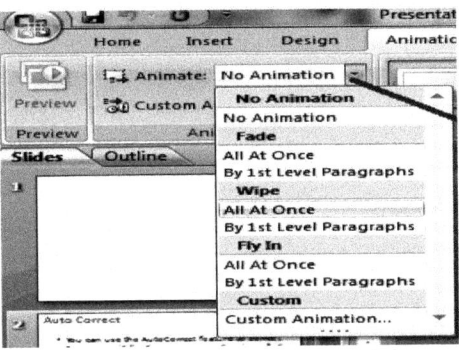

On the other hand, "custom animation" allows the user to set the effect as per their choice of available options. To show the effects on the slides, the following steps are to be taken. Select the desired text (Step: 1) on the slide to which an effect is to be added, click on the "Custom Animation" (Step: 2) button, another menu opens on the right hand side of the screen. In it click on the "Add Effect" button (Step: 3), another menu opens up, in it select the category (Step 4) and in it select the desired effect (Step: 5) to be applied on the selected text as shown in the below figure:

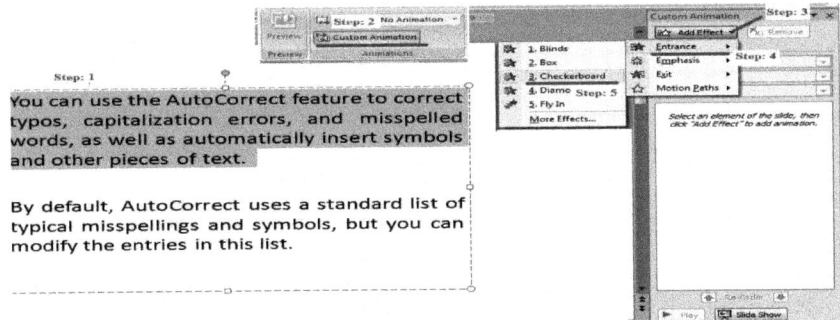

The speed of the text animation effect can be controlled by selecting the desired option as shown in the following figure:

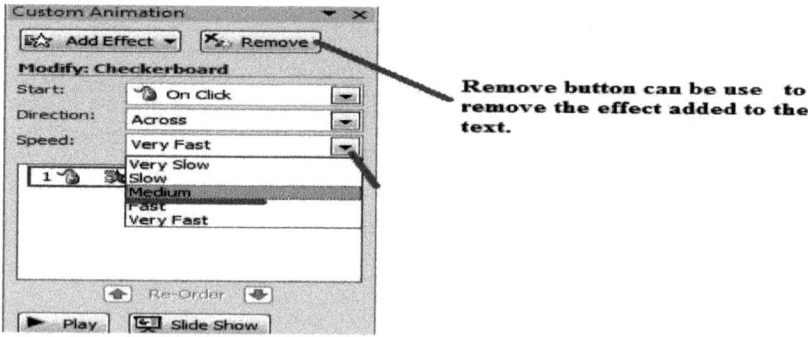

3.4.5 Slideshow Tab

Clicking on the "from beginning" button will start the slide show from the first slide, whereas clicking on the "from current slide" button will start the show from the present slide.

The "custom slide show" button allows the user to reposition the slides and see the show according to the new sequence. When the user clicks on the "custom slide show" button, a menu will appear and the user should click on the "new" button. This will cause another screen to appear and in it reposition the slides and click "OK."

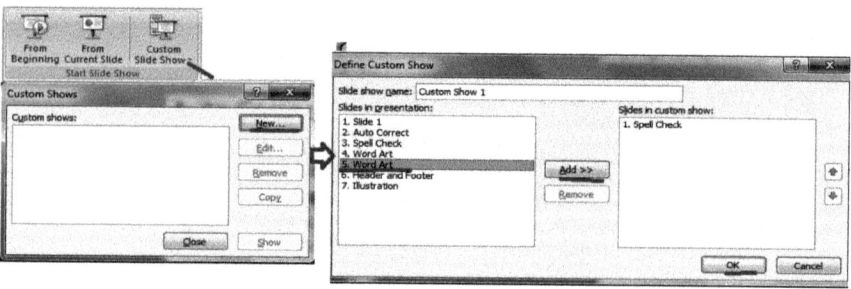

A window will open (as shown in the next figure) with a new option of "custom show 1. "Clicking on the "show" button will start the show as per the new sequence set.

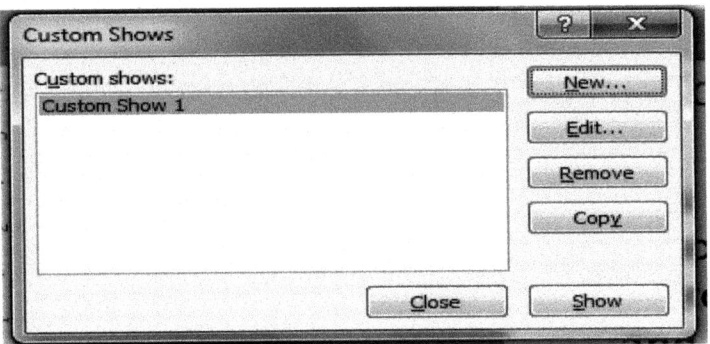

3.4.6 View Tab

PowerPoint provides the user with a number of views.

Normal View: This is the default view. It is the view in which a user writes or designs his presentation. Normal view has four working areas:

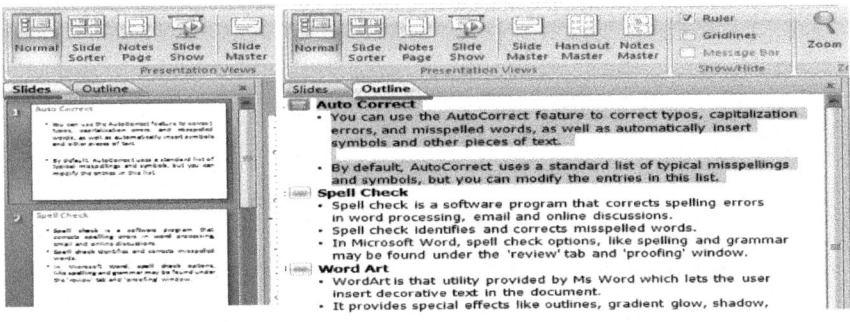

- **Slide Tab:** This is a thumbnail-sized image of the slide that appears while we create it. The thumbnails make it easy to move through the slides. This section also allows the user to rearrange, add, or delete slides here.

- **Outline Tab:** The outline tab shows the slide text in outline form where it can be edited.

- **Slide Pane:** This is the area where the user prepares the slides by default in a normal view. The current slide is always shown in this view. The slide can be edited and in the user can insert pictures, tables, charts, movies, sounds, hyperlinks, and animations.

- **Notes Pane:** This is below the slide pane. Here the user can type notes related to the current slide.

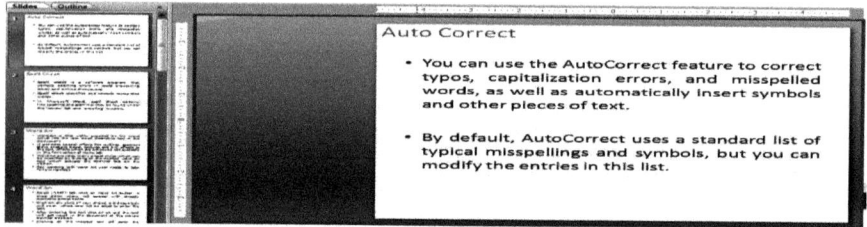

Slide Sorter View: This view provides all the slides in thumbnail form. It basically helps the user to arrange the sequence of the slides.

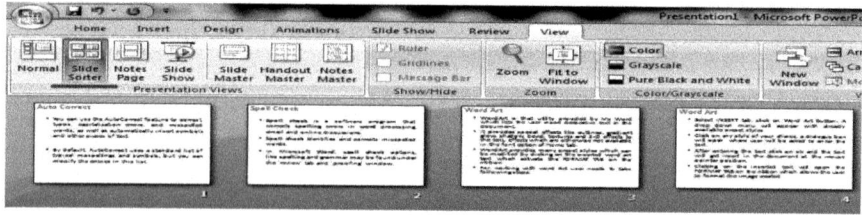

Notes Page View: The notes pane is located just below the slide pane. This place is used by the user to type notes related to the current slide. In order to view and work with notes in a full page, the user needs to click on the "view tab" and under it click on the "notes page" button.

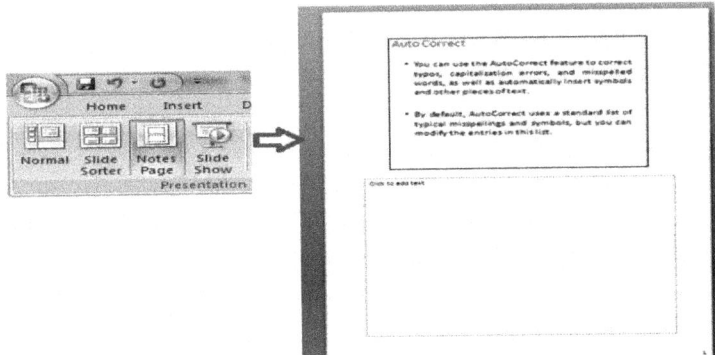

Slide Show View: Clicking on this will start the slide show.

3.4.7 PowerPoint Show

Saving a PowerPoint presentation as a PowerPoint show helps the user to start the presentation automatically when the file is opened. It is saved as a .ppxs file. Click on the Microsoft Office button, and in it click on "save as" to select the option "PowerPoint Show" as shown in the following figure:

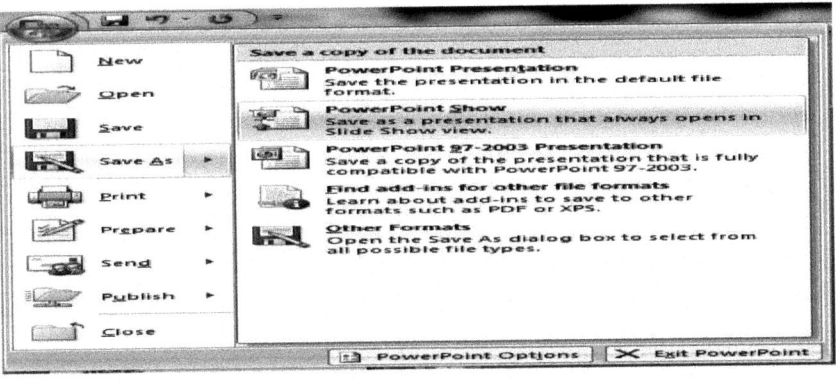

3.5   Database Management System Using Microsoft Access

3.5.1 File Terminologies

The following are some various terminologies used in the design of a file.

Data: Data is a collection of facts, figures, and statistics that can be processed to produce meaningful information. For example, names of students, marks, and so on. The smallest unit of data is a character and there are three types:

- Alphabetic : a—z and A—Z
- Numeric : 0 to 9
- Special : !, #, $ * = etc.

Individual items of data are called data elements. For example, details about a student and their marks, i.e., name, class, age, marks 1, marks 2, etc.

Field: When data is stored in a file, they are called fields. Like data elements, fields too have names to which they are referred. For example, name, age, class, marks 1, marks 2, etc.

Record: A group of logically related data items or fields is called a record. For example we have a number of fields like name, age, class, marks 1, marks 2, total, etc., and all these fields put together in relation to a student will form a record.

Database File: Thus a file can be defined as a collection of related records. For example, details of one student such as name, age, class, marks 1, marks 2, total, etc. is a record, and if there are 50 students in a class, each will have its unique record, and a collection of these 50 records will be termed as file.

3.5.2 Types of Data Files

- **Master File:** A master file is a file of relatively permanent nature. In it information about entities stays unchanged for a long period of time. Master files are used as an original source of data for processing transactions and generating information based on the processed data. For example, the master file of accounts data in a bank will contain details like account number, type of account, name of the account holder, address, date of opening the account, nominee, balance details, and so on.

- **Transaction File:** When data is to be processed for a required purpose, a temporary copy of the master file is copied into the memory of the computer system where the result of processed transactions is stored and transaction documents are prepared. This is a collection of records describing activities or transactions by an organization and is known as a transaction file. Later, the data from the transaction files are used to update the details in the master file. For example in banks, the transaction file will contain records of the day-to-day activity of the bank, that is, deposits, withdrawals, transfers, and so on. At the end of the day, the transaction file is processed and the master file is updated to reflect the changes that had occurred during that particular day.

- ▪ **Report File:** This is a file created by extracting data to prepare reports. A report of all accounts sorted by the account number and containing the details like account holder's name, account balance, etc. is an example of a report file.

3.5.3 Types of Records

Fixed-Length Record: When the number and size of the data element in a record is constant, then the record is called a fixed-length record. For example:

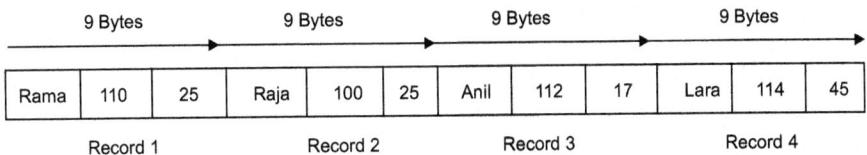

In a fixed-length record, an optimum amount of memory is allocated for every data element. For example, if we define name = 20, it means 20 bytes of memory is allocated to the data-field name in each record. This size does not vary, i.e., if we store the word "Anil," it will occupy 20 bytes, although it contains only 4 characters. On the other hand if we try to store a name of more than 20 characters, only the first 20 characters will be stored in the data field and the remainder will be truncated. But by fixing the length of the record, the processing speed becomes fast and the computer knows exactly where a new record begins.

Variable-Length Record: If different records in a file have different sizes, the file is said to be made up of variable-length records. In a variable-length record, only the amount of storage space required by the data is allocated at the time of entering the record. The size of the record varies as the size of the individual data element will vary. Storage space is saved at the cost of processing speed as a system needs to know where a particular record begins, where it ends, and the size of the record.

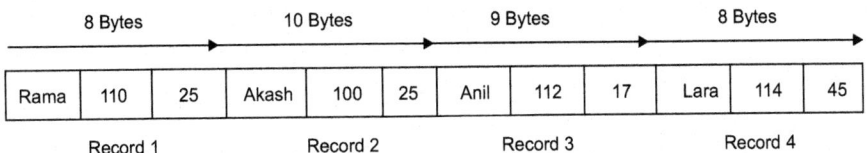

For example, in an organization some employees may be taking one insurance policy, some may be taking two or three. If these were to be stored in a fixed-length record, then it is required to know the maximum number of policies that exist and to allocate the memory space accordingly for each employee record. Obviously this would amount to a waste of space, as not all employees will go for all of the available policies. Thus, the best way of storing is to have a variable-length record and to allocate space that is required for the actual number of policies taken by an employee. Although this saves space, it makes processing much more complicated and hence increases the processing time.

3.5.4 Database Management System

A database system is a combination of software and hardware that makes it possible and convenient to perform one or more tasks that involve handling large amounts of data. The four major components of a database system are:

- data
- hardware
- software
- user

The objective behind this creation is to make information access easy, quick, inexpensive, and flexible for the user. A DBMS acts as an interface between the user of the system and the database, which controls the database and performs the entire job required by the user on the database.

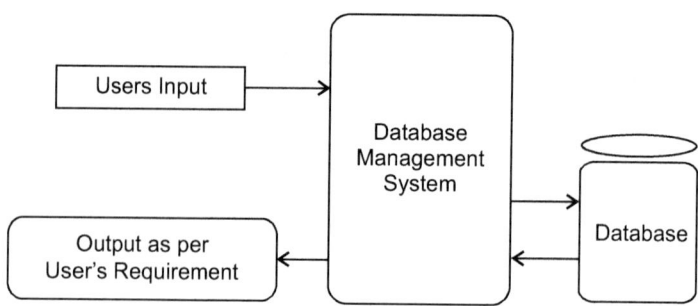

Once the groundwork is done, then the following steps are required to be taken to create a data file in MS Access. Start the MS Access software by taking the following steps:

- Click on the start button.

- Under program, select Microsoft Access.

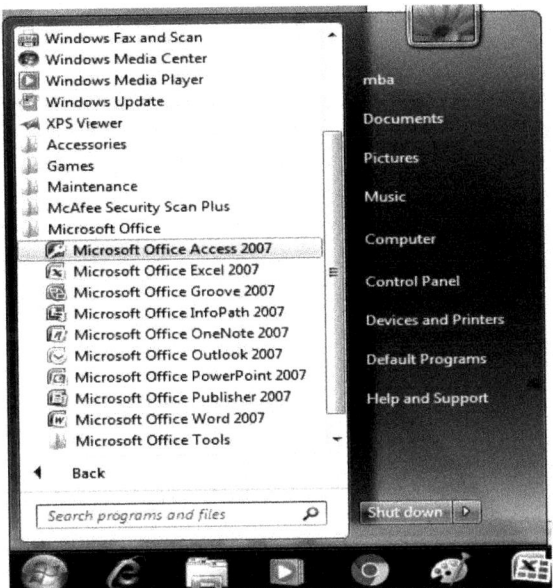

- A new screen will appear as shown in the following:

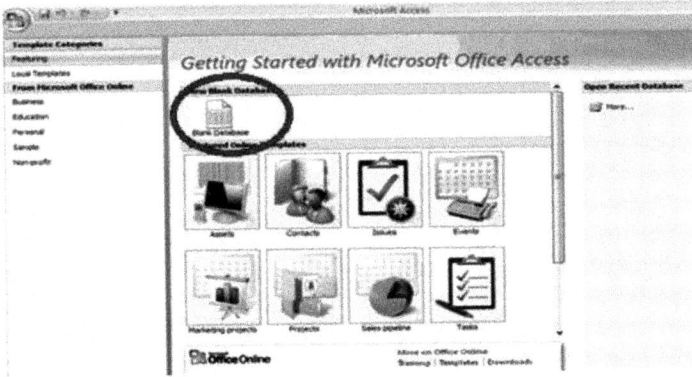

Click on the "blank database" option. A new page opens up on the extreme right-hand side asking the user to enter the location and a database file name. Type the name as new and click on the create button.

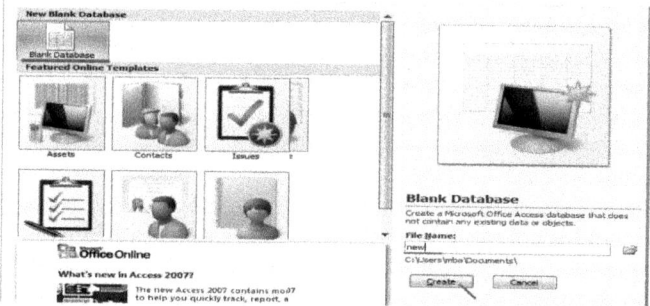

A new screen will appear and is known as the "datasheet view." Here the values can be entered, but first we need to design the structure of the data file, so click on the "view" option and select "design view" to design the structure of the database.

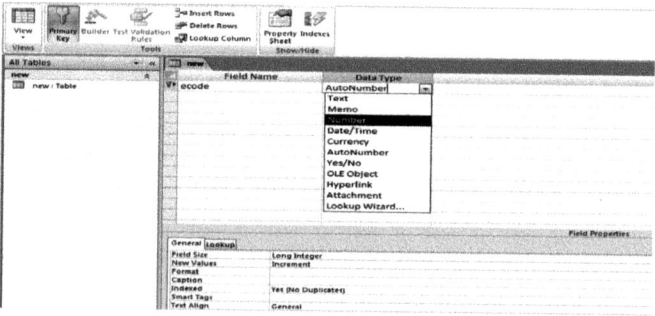

Here the field names are the attributes of an entity which a user needs to define.

3.5.11 Data Type

Data type basically specifies the nature of the data that field is going to store. MS Access allows the use of following data types:

Data Type	Description
AutoNumber	An AutoNumber field creates unique values automatically when Access creates a new record. The AutoNumber field is primarily used for primary keys in Access.

Text	A text field can contains values that are text, numeric, or a combination of both. A text field can contain a maximum length of 255 characters.
Memo	A much larger version of the text field, allowing storage of up to 2 GB of data. A new feature of the Microsoft Access Memo field is that it now supports rich text formatting.
Number	The number field can store numeric values up to 16 bytes of data.
Date/Time	The date/time field allows storage of date and time information. The date/time field now also includes the auto calendar feature.
Currency	The currency data type stores values in a monetary format. This can be used with financial data as 8-byte numbers with precision to four decimal places.
Yes/No	Boolean data storage of true/false values.
OLE Object	The OLE object field stores images, documents, graphs, etc. from Office- and Windows-based programs. The maximum data size is 2 GB, although this will slow down a database.
Hyperlink	The hyperlink field type is used to store web addresses. This has a maximum size limit of 1 GB of data.
Attachment	The attachment field type is used to store images, spreadsheet files, documents, charts, and other types of supported files to the records in your database. This is a feature that was introduced in Microsoft Access 2007.

3.5.12 Elements of a Database

The database window gives users the following options and can be created by clicking on the respective button:

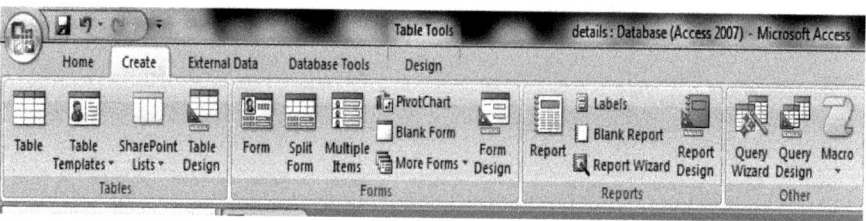

1	Table	This is the place where the fields are specified and then data is stored for further processing. Number of tables can be created and can be linked to each other called defining a relationship among the data files.
2	Forms	Forms are created to have a simplified display of the data where the user can also make required changes, enter new values, and can display it on the screen using the layouts available in the Form option.
3	Reports	Reports are prepared for the purpose of presenting the processed data in front of the user or the people who are interested in it. Reports can be based on the group data or on any other specified conditions.
4	Query	Whenever the user wants to perform any operation on the data file like displaying records based on some conditions or in general, Query option is used. A query can also modify multiple records at the same time.
5	Macros	Macros are like a button pressing of which initiated a series of processes. With a single command many database tasks can be automated.

3.5.13 Primary Key/Record Key/Unique Key

Once the field names and their respective data types are specified, the next step is to specify the k "primary key" or a unique Key (also known as a record key). A "primary key" is that field of the table that cannot be the same for any two records. For example, in our "employee" data table, "Ecode" which is the employee code, is going to be the "primary key" as in an organization no two employees can have the same employee code. To set a field as a "primary key," select the field by pressing the left mouse button and then press the right mouse button. The following screen will appear:

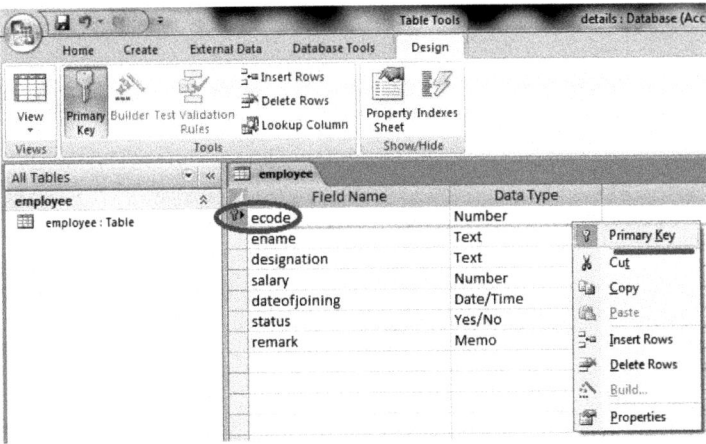

Click on the primary key option and the field Ecode will be set as the primary key for the data file. A "primary key" is set so that a relation can be set among the different data files or tables of the organization consisting of data.

Data Entry: For the purpose of entering data in the table simply click on the "view" button on the "home" tab and click on the "datasheet" view. Another screen will appear in which the data can be entered and will give the following display:

Once the data in the last field is entered, which in our case is remark, pressing the tab key will move the cursor automatically to the next record and access will save the previous record automatically. After the table has been created, if the user needs to make the changes in the table structure, for example in the field name or in the data type, the following steps are to be taken:

1. In the database window click on the "design" option and the field view will appear. This is where the required changes can be made.

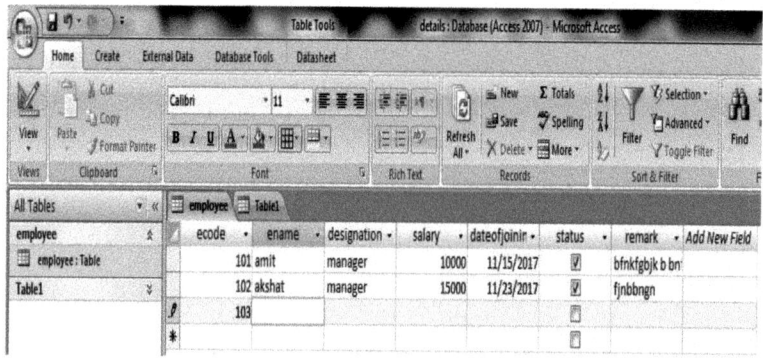

2. If a user needs to add a new field, this can be done in both the datasheet view as well as design view. Place the mouse pointer at the "header column" in the "datasheet view" or on any row under the "field name column" in "design view." Now right click on the mouse and select the insert column or the insert row option.

3. In the datasheet view after inserting the column the user needs to rename the column, i.e., needs to give it a field name. This can be done by right clicking on the column inserted and selecting the "rename column" option.

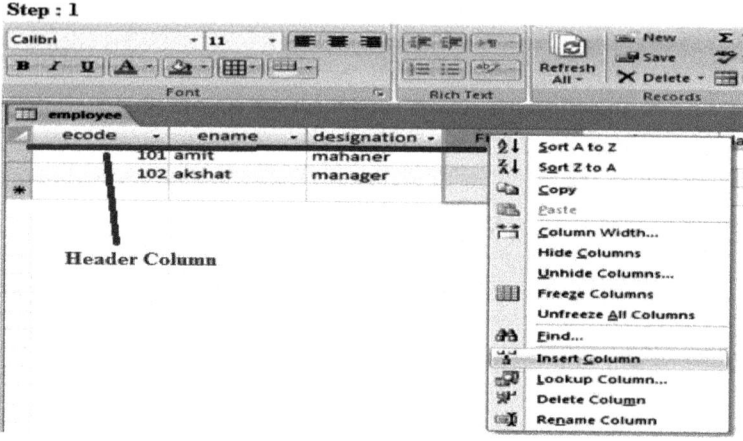

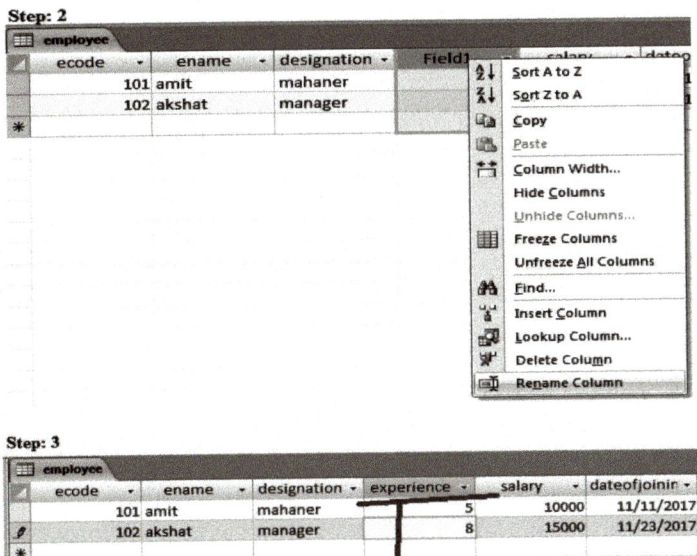

Now the table is ready to be populated with data. To access and to display the data as per the user's requirements, queries, forms, reports and other options can be used.

Test Your Knowledge

1. What is MS-Office and for what purpose can it be used?

2. Explain the utility of the status, horizontal, and vertical scroll bars in MS-Word.

3. What do you understand about the nonprinting characters in MS-Word? Explain.

4. Explain the functioning of the following keys in MS-Word.

- backspace key
- delete

- insert text
- Bold

- underline
- Italicize

5. Explain the following options under the Office button in MS Word: New, Open, Save, Save As, Print, Prepare, Send, Publish, Close.

6. Explain the utility of the spell check and grammar option in MS-Word.

7. Explain the concept of bullets and numbering.

8. Explain the difference in font size and font type options under the home tab.

9. How you will create a table using MS-Word? Explain.

10. Explain how the printing of a document can be done in MS-Word.

11. Explain the use of ClipArt, inserting the picture from file, and the Word Art in MS-Word.

12. What is alignment? Explain the difference in text alignment and cell alignment.

13. Define clipboard.

14. What is hyperlink? Write down the steps to create a hyperlink.

15. What is header and footer? How are they created? What is the utility of having a header and footer in a document?

16. What is page orientation? How many types of page orientation are there?

17. What are margins? How will you set the margins?

18. Write down the steps to split the text into two or more columns.

19. Explain watermark.

20. What is an indent? How it is different from a hanging indent?

21. What is thesaurus? How it is useful?

22. How will you change the page orientation?

23. Explain the process of mail merge in MS Word.

24. Define macros. Write down steps to create a macro.

25. Explain the various tool bars of Excel.

26. What do you understand about a worksheet in MS Excel? What is the use of a formula bar in Excel?

27. Explain the use of the tab key, page up key, page down key, end key, and home key in Excel.

28. How is data entered and edited in Excel and how are the cells are selected?

29. How can the width of a cell and the height of a row be changed in Excel?

30. Explain the process of cell alignment, fonts, border, and pattern.

31. How can a user select a new sheet in a workbook?

32. How can mathematical calculations be performed in Excel? Explain.

33. Explain how the numbers can be formatted in Excel?

34. What do you understand about cell addressing in Excel? How many types of cell entries can be there in Excel?

35. What do you understand about a function in excel? Explain the use of a few commonly used functions in Excel.

36. What are the various categories of functions used in Excel?

37. What do you understand about user-defined functions? Explain.

38. How can a cell be filled automatically in Excel?

39. What is a chart? How is a chart prepared in Excel? Discuss.

40. What are the various types of charts available in Excel? Discuss.

41. Assume sales figures of four salesmen for all quarters and write down the steps to draw a column chart to present the data graphically.

42. What do you understand about the sort command in Excel? Explain it with the help of an example.

43. What is a filter? How are filters created? Explain with the help of an example.

44. What do you understand about PowerPoint? What are its various features?

45. How do you set up a template presentation page in PowerPoint?

46. What do you understand about "view" in PowerPoint? How many types of views are there?

47. For what purpose is the option slide master used in PowerPoint? Explain.

48. What do you understand about a slide layout?

49. What is a slide transition? How it is different from text animation?

50. What is a PowerPoint show? How it is created?

51. Explain the following terms:
  (i)   data
  (ii)   field
  (iii)   record
  (iv)   file

52. How many types of data files are there? Explain their differences.

53. Explain the difference between a fixed-length record and a variable-length record.

54. What do you understand about a database management system? What are its properties?

55. What are the various advantages of a database management system?

56. Discuss the various applications/usages of a database management system.

57. Write down the steps to create a table in MS Access.

58. What do you understand about data type? How many data types are there in MS Access?

59. Define the process of creating a table, form, or a query in MS Access.

60. What do you understand about a primary key, record key, and unique key?

CHAPTER 4

MANAGEMENT INFORMATION SYSTEMS

4.1   Need for Data and Information

In today's world, business organizations are facing a cut-throat competition in the market place. It has become very difficult to survive, and it is becoming harder to maintain the market and the market share. It is only with proper access to data and the information generated from that data, that can help business organizations in making quick and relevant decisions. These decisions not only help business organizations in retaining their market shares, but also helps them in keeping proper track of their competitors' activities in the market place.

In today's business organizations, data and information are no longer treated as mere tools for conducting business, rather they are considered important assets that help them in making proper and timely decisions at various levels of management, such as decision support systems at the middle level of management, and executive support systems at the top level of management. These decisions process data to generate information used for various business purposes.

4.2   Levels of Information

In a business organization, generally three levels of management are found. They are:

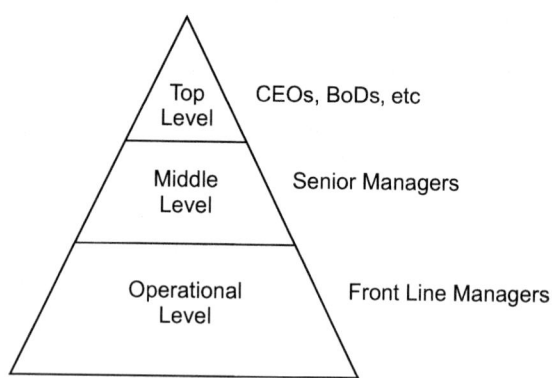

Each of the three levels are involved in various business processes. Top level or the strategic level of management makes decisions on issues such as launching a new product into the market, deciding on the financial issues of the business, etc. Middle or the tactical level of management is concerned with the process of ensuring that the decisions taken by the top or strategic level of management are properly implemented, and the resources required to perform various jobs are made available to the business organization. The operational level or the front-line level of management is concerned with the actual implementation. Thus, based on the level of management, the information itself can also be divided into three categories.

- **Strategic Information:** This is the information needed for long-range planning and deciding the direction the business should take. Strategic information is used by the top management to make plans for the organization and to ensure that the business objectives are achieved with the help of a system called an *executive support system.*

- **Tactical Information:** Tactical information is used in making short-range decisions to run the business efficiently. Tactical information is used by middle management for ensuring that the resources are used properly to achieve what is decided by the strategic management. Tactical information is usually used for decision making at the middle-management level with the help of a system called a *decision support system*.

- **Operational Information:** Operational information is required for the smooth conduct of the daily operations of a business organization. Operational or front-line managers use the operational information to ensure that assigned tasks are planned and carried out properly in the business organization.

- **Levels of Management and Related Information Need:** The structure of an information system can be classified in terms of a hierarchy of management planning and control activities:

Information Type	Example	Level of Management
Strategic	Should a new branch be opened? Should the business be diversified?	GEO, Directir
Tactical	How to rate vendors How much of each item should be stocked? Should new discount policy be introduced?	Middle Management
Operational	List of items to be reordered List of defaulting customers Daily ledger accounts List of outstanding bills to be paid	Line Managers

4.3 Characteristics and Presentation Qualities of Information

- **Quality of Information:** The quality of information with respect to the decision maker depends on two factors. One is the utility of information, and the second is based on the satisfaction level generated by that information. These can be defined as follows:

- **Utility of Information:** Information must be evaluated in terms of its utility, including the accuracy of information that might facilitate or retard its use. Following are the three information utilities:

  - **Form Utility:** If the information is provided in a format that is based on the need of the decision maker, its value increases.

  - **Time Utility:** If information is provided when it is needed, it is of greater value to the decision maker and thus enhances the utility of information.

  - **Place Utility:** If information is delivered at the right place or is easily accessible, it has a greater value.

- **Information Satisfaction:** If the preceding three utilities are met, the decision maker will be satisfied with the output of the information system, and this will determine the level of quality of the information.

Characteristic of Information

Accurate	Ensure correct input and processing rules
Complete	Should include all data
Trustworthy	Do not hide unpleasant information
Timely	Should be given at the right time
Brief	Summarize relevant information
Significantly Understandable	Use attractive formats and graphical charts

4.4 Value of Information

At any given time there are a number of options available to a decision maker, and those options will select the best ones based on the information that is currently available. But before a decision is made, if new information becomes available and it causes a different decision to be made, the value of the new information is the difference in the value between the outcome of the old decision and the new decision. If the new information does not cause a different decision to be made, the value of the information is zero. It can be represented as:

Value of Information = Value of outcome (new decision) –value of outcome (old decision)

For example, there are three alternatives: X,Y, and Z. The decision maker on the basis of prior knowledge (in "perfect knowledge" values) estimates that the outcome from X = 20, from Y= 30, and from Z =15. Therefore select Y as the "new decision perfect information" that is provided, which establishes without doubt that the payment from Z equals 30 and from Y it is only 22. This information causes the decision maker to select Z instead of Y, thereby increasing the payment from 22 to 30. Thus the value of perfect information is therefore 8:

Payment List 1	Payment List 2
X = 20	X = 20
Y = 30 Decision = Y	Y = 22 Decision = Z
Z = 15	Z = 30

4.5   Information System

An *information system* can be defined as a system that receives the data and instruction as input, processes data as per instructions, and generates the output known as information. An information system is a combination of users, computer hardware, computer software, communication networks, and a data resource that collects, transforms, and disseminates information in an organization. Today there are various types of information systems available in an organization, and its proper management is a major challenge for managers. An information system typically performs three vital roles in any type of an organization:

- supports business operations

- supports managerial decision making

- supports strategic competitive advantage

Structure of an Information System: The information system of an organization is typically made of its physical components, processing functions, and output to the users as defined here:

- **Physical Components:** The physical components of an information system consist of hardware, software, database, procedures, and people operating these components.

- **Processing Functions:** Following are the functions that are performed by an information system in an organization:

Processing Functions	Activities Performed
Process Transaction	A transaction is an activity such as making a purchase of sale or manufacturing a product. It may be internal to the organization or external. For example, the process of booking an airline ticket is a transaction.
Maintain Files	It is very common in an organization to create and maintain files of every transaction that takes place. These files store data that remains relatively permanent in nature. For example, an airline reservation system maintains the record of each passenger, the details prior to the trip and even after the trip (hotel, auto rental, etc.. When transactions are processed, appropriate files are updated with the latest information.
Produce Reports	An important output of an information system is the various types of reports produced by it. Scheduled reports are produced on a regular basis. An information system should also be able to produce special reports quickly on an *ad hoc* basis.
Process Inquiries	Another important function performed by an information system is to process inquiries using the data stored in the database. The essential function of inquiry processing is to make any record or item in the database easily accessible. For example, a passenger inquiring about the ticket status from the airline reservation system.

- **Output to the user:** The output can be classified by the following types:

  - **Status report:** This report produces the current status of any ongoing activity.

  - **Confirmation of previous knowledge:** This could have two possible results:

  - **Correction of previous knowledge:** For example, the passenger's ticket is on a waitlist and now the status is confirmed; so the previous knowledge now stands corrected.

  - **Surprise elements:** The passenger is surprised to know that the ticket is not yet confirmed.

- **Analysis:** Generated reports help an organization in doing many types of analysis. For example, reports are generated about performance analysis, sales analysis, etc.

- **Predictions and forecasts:** On the basis of the analysis performed, organizations can make predictions and forecast for future developments.

- **Optimizations:** Decision making is always a complicated process because organizations frequently operate in a very dynamic environment. There are a number of parameters that vary. Some of these are independent and others are not. A manager needs to evaluate the complete effect of a decision and make tradeoffs where necessary.

- **Frequency:** This means the time period in which reports are produced. Frequency of the reports can be classified as follows:

  - **Regular reports:** The routine reports for status reporting (e.g., monthly).

  - **On-demand reports:** These are to be produced whenever someone asks for them.

  - **Exception reports:** Certain reports are only required when exceptional conditions are reached to highlight special cases.

  - **Ad hoc reports:** These are more or less unplanned. One would not typically have programs for these as part of an information handling system. The user would need to design the report with tools like a report generator or by requesting a special service from the EDP department.

The information system function nowadays is considered as a major functional area of business and it is equally important for the business success as other functions such as marketing, sales, finance, production, and HR. It is also considered as an important contributor to operational efficiency, employee productivity, morale, customer services, and satisfaction. It acts as a major source of providing information to the management which leads to effective decision making. However, it also poses a big challenge as it requires a high level of investment on a continuous basis.

4.6   Components of an Information System

An information system consists of five major components. They are *users, hardware, software, data,* and *networks.* These components perform processing that converts data into meaningful information products. The following diagram shows a framework of all of the major components and of information systems.

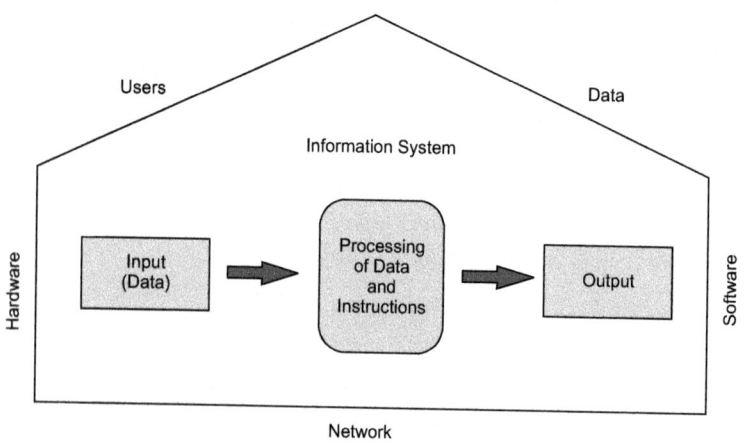

4.7   Types of Information Systems

There are a number of ways in which an organization's information systems can be classified. However, all of the systems working in an organization can be broadly classified into two categories: a) One supports the routine office and operations work and b) the other helps management in the decision-making process. The categorization is shown in the following figure:

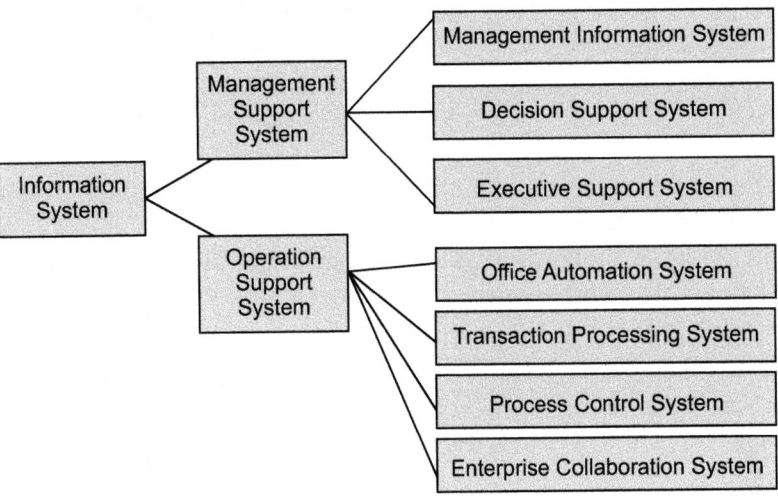

4.7.1 Operation Support Systems

These are the systems that help the business organizations perform routine processes. These systems produce information that helps in the automation of routine office work, efficiently processing business transactions, controlling production processes, and facilitating the communication and collaboration among members of the organization. The systems falling under the category of the operating support system are described in the following section.

4.7.1.1 Office Automation Systems

Any office is expected to perform a number of tasks known as routine office tasks. Such tasks include drafting letters, tracking schedules, tabulating data, and many more. There are a number of packages available to help employees accomplish these routine tasks. Some examples are MS Office, communication software, etc. Some of the routine tasks performed by an office automation system are as described as follows:

- **Filing:** Computer-based filing systems save space and reduce the need for maintaining essential documents in paper format.

- **Using Graphics:** It is a well-known fact that if the data or text can be presented in graphical format, its utility increases. The office automation system provides a number of graphical tools for this purpose.

- **Communication Networks:** Communication networks are the networks by which communication flows. In an office all of the systems are connected to one another with a central database that provides data and information to all of the work stations as needed.

- **Computer-Based Message Systems:** Video conferencing capabilities allow people to communicate face-to-face without traveling. Such capabilities were initially used to connect a few executives in two or three locations. These days, such systems are used to connect thousands of people to the same presentation. A private branch exchange (PBX) allows for local office communication over telephone lines. A branch exchange is a switching system that enables telephones to be connected to each other.

4.7.1.2 Transaction Processing Systems (TPS)

Transaction processing systems (TPS) are used at the operational level in an organization. These systems record the elementary activities and transactions such as sales, receipts, cash deposits, payroll, the flow of materials, and so on. Information must be accurate, updated, and easily available. These systems:

- are the connection between an organization and its environment (internal and external), and

- provide data and information to other systems in an organization.

There are typically five types of TPS in an organization. They are *marketing, production, financial, personnel,* and *industry specific.*

- **Transaction Processing Cycle:** A transaction is recorded on a transaction document. These may be on hand-written forms or through an online terminal. Master files are then updated through batch or online processing. Input records are validated to determine if they are correct and complete. Validation may be for size, range, and composition of characters. During processing the master file is updated and outputs such as transaction documents and reports may be generated. Transaction data output may be for:

  - **Information:** to report, confirm, or explain a proposed or a completed action.

  - **Action:** to direct a transaction to take place.

  - **Investigation:** background information or reference.

Action documents instruct someone to do something. A purchase order instructs a vendor to dispatch, a check instructs a bank to pay, etc. During processing, a listing of data about each transaction is usually prepared. The listing represents a batch of transactions or in the case of online processing, the processing during a period of time.

■ **Processing Transactions:** Transactions are processed through:

    (i) **Online Processing:** This means data is processed at the time of the transaction taking place.

    (ii) **Batch Processing:** This means data is first collected and then it is entered into the system in a single session.

4.7.1.3 Process Control Systems

Process control systems are those systems in which decisions adjusting a physical production process are automatically made by computers. For example, in a refinery, electronic sensors which are linked to computers are used to monitor the processes on a continuous basis. The computer captures and processes the data detected by sensors, and makes any instant adjustments if deemed fit to the refinery processes. Another example is that of Frito Lays potato chips production unit, where a laser is thrown on each chip, and the chips that have the slightest burn mark are popped out from the production process to maintain the quality standard of every single chip.

4.7.1.4 Enterprise Collaboration Systems

Time is costly and today's organizations understand that very well. The rise of the Internet and its utility in performing business operations has dramatically changed the computing mentality of organizations. The objective of enterprise collaboration systems is to help users to work together efficiently and effectively by providing help with:

■ **Communication:** This is the sharing of information with one another.

■ **Coordination:** Coordinating individual work efforts and use of resources across the organization.

■ **Collaboration:** Collaboration is working with one another cooperatively on joint projects and assignments.

    (i) **Workgroups, Teams, and Collaboration:** When two or more people, known or unknown to one another, work together on the same task, it is called a *workgroup*. On the other hand, when this group collaborates on joint projects, it means they work in harmony and they are called a *team*. Thus one can say that working with one another in a cooperative way toward the achievement of a common goal is called *collaboration*. Collaboration is that fundamental concept that is a key to success for an organization. It is that concept that makes a group of people a team, and what makes a team successful. It is not at all essential that members of a team or workgroup need to work at the same location. They can be at different locations in a city or a country or in different continents. *Enterprise collaboration systems* make working from different social fields possible. This system allows end users to work with one another in virtual teams without regard to time constraints, physical location, or organizational boundaries.

(ii) **Enterprise Collaboration System Components:** An Enterprise Collaboration system is basically an information system with different objectives. Thus, like other systems its components include hardware, software, data, and network resources. These components of the system are used for the purpose of communication, coordination, and collaboration among the members of team. For example, a software company takes a project in the United States. Now the company may form different teams to work on the project. One in Bangalore, one in Mumbai, and one in the United States. This team will make use of intranets and extranets to collaborate via e-mail, ZOOM, Skype, MS Teams, and a work-in-progress information at a project web site.

(iii) **Groupware for Enterprise Collaboration:** *Groupware* is defined as software that helps in collaboration. For example, Smartsheet, Microsoft Office, Oracle Beehive, etc. support collaboration through e-mail, data, and audio conferencing, discussion forums, scheduling, and calendaring. Groupware software makes it easy for the system user to communicate effectively; coordinate workgroup activities within the time limits allowed irrespective of the physical location of the members of a team.

(iv) **Electronic Communication Tools:** These are e-mail, voice mail, web-publishing, bulletin board systems, paging, and Internet phone systems.

(v) **Electronic Conferencing Tools:** Members of teams, at the same or different locations uses numerous conferencing tools for exchanging their ideas and to interact simultaneously, or at different times as per need. To achieve this, the team needs conferencing tools communication. Various conferencing tools used by the team for this purpose are data and voice conferencing, videoconferencing, teleconferencing, discussion forums, chat systems, animation house, web forum, electronic meeting systems, etc.

(vi) Collaborative Work Management Tools: It is essential for a team to work in a time frame so that they are able to complete tasks as per the set deadlines. Collaborative work management tools help teams in the management of all work activities. Calendaring and scheduling, project management, and knowledge management are a few of the tools that help in the work management.

4.8   Management Support Systems

4.8.1 Management Information Systems

Various definitions provided by different authors for MIS are summarized as follows:

- The MIS is defined as a system that provides information support for decision making in the organization.

- The MIS is defined as an integrated system of man and machine for providing the information to support the operations, the management, and the decision-making function in the organization.

- The MIS is defined as a system based on the database of the organization evolved for the purpose of providing information to the people in the organization.

- The MIS is defined as a computer-based information system.

If one analyzes these definitions, one will find there is a single focus on the definition of MIS: the MIS is a system that supports the decision making in an organization. Thus it can be said that *MIS is a computerized data-processing system that generates information for the people working in the organization to meet their information needs for decision making.* Management information systems are designed to support the need of three different categories of managers: top level, tactical level, operational level, and it also has following applications:

- It supports structured and unstructured decisions

- It is generally reporting and control oriented. They are designed to report on existing operations and therefore to provide help in day-to-day control of operations.

- It performs different types of analysis.

4.8.2 Decision Support System

A system which supports the middle (tactical) level and also the higher level of management in arriving at a decision is termed as a *decision support system* (DSS). DSS is a system that reduces the uncertainty surrounding the decision-making process and helps managers in making decisions that are not easily specified in advance. DSS not only provides relevant information to the management, but it also suggests the possible decisions to be made in a particular situation. DSS are capable of running several times a day in order to respond to the changing business environment. DSS have more analytical power than any other systems working in an organization. A few of the important functions of a DSS are:

- DSS offers its users flexibility and prompt responses.

- DSS lets the users have control on the input and output of the process.

- To work on DSS, managers need not have a specialized computer programming knowledge.

- DSS uses sophisticated analysis and modeling tools for generating results and are knowledge-based systems.

4.8.3 Executive Support System

The top strategic level of management is often confronted with situations that are unique in nature and are not repeated. *Executive support systems* (ESS) are systems that help the top level of management in the decision-making process. ESS incorporates many features of the MIS and DSS, and also helps the strategic level by providing information so that they can come to a conclusion. ESS does not provide top management with any fixed set of solutions, but rather it creates a generalized computing and communication environment that can be applied to many situations that demand decisions. The designing of ESS is usually done in such a manner that it incorporates data about external events, such as changes in tax policies or new competitors in the market place. ESS employs tools to compress and filter critical data out of piles of data, which helps in saving time and efforts required to obtain information which is useful to the management for decision making. ESS provides advanced graphics tools that help in presenting data in graphical form making it easy for senior executives to comprehend. The graphical display of data allows senior executives to reach down to the lowest possible detail in no time and thus reduce the uncertainty surrounding the decision. Thus one can say that in ESS the processed data or information is presented in front of the senior executives in a manner that facilitates fast decision making.

4.8.4 Expert Systems

Experts are the people who possess specialized subject knowledge like accountants, architects, electricians, etc. Similarly, when an organization faces a problem that is very different and specialized in nature, they need help and advice from an expert. The expert is someone who possesses required knowledge and experience in a particular area, and is fully aware of alternatives, chances of success, and costs and benefits to the organization. Thus it can be said that if it is a problem that is structured in nature, organizations use computer-based expert systems.

Expert systems are based on *artificial intelligence* technology and are the most popular and widely used applications in business. They are knowledge-based information systems and use expert knowledge in a specific domain in the business decision-making process. Expert systems work just like a human expert who provides solutions to problems in defined and specific areas. Solutions provided by expert systems might be in the fields of taxation, financial analysis, maintenance of equipment, fault diagnosis, etc. They basically act as tools that help in the improvement of the productivity and quality of decision making.

It is inferred that expert systems have a limited knowledge. They are designed to perform very limited tasks that can easily be executed by a human expert. For example, calculating tax on income, diagnosing a malfunctioning machine, or the checking of a credit score. However, the expert systems designed today are flexible and have a large knowledge bases and reasoning methods. For this they make use of neural networks with rule-based inferences to achieve a higher decision performance.

Components: Every expert system is supposed to have three basic components—*knowledge base, inference engine, and user interface. If-then-else rules* are used by the expert system to represent and store knowledge, and they are capable of applying reasoning methods to solve a problem. *Knowledge base* is basically the set of rules that are required to solve any problem. For example, tax rules for calculating tax of an individual, organization, or a senior citizen. These rules are written in the software program for the purpose of decision making. The process of searching the knowledge base for an applicable rule is called an *inference engine*. It's like Google used by people to search required information on the Internet. The inference engine is basically a software program. *User interface* is basically a user who uses an expert system to find a solution to his problem.

Applications: Expert system applications are used extensively in business. Benefits accruing to businesses because of the use of expert systems range from better decisions, minimized errors, and cost reductions to the improved levels of quality and service. Its knowledge-base helps organizations in assessing the risk. They are used by organizations in customer relationship management, human resource management, manufacturing, dealer evaluation, marketing and sales, etc. For example, Logitech, one of the world's largest manufacturers of mouse and web cameras, uses expert systems for customer support. The web-based software emulates the way a human would interact with a customer, allows the user to ask questions or describe problems in the natural language, and carries on an intelligent conversation with the user and provides an accurate answer.

Worldwide there are web-based, self-assessment and income tax return filing systems, where the individual can enter the required information and the system calculates tax, then generates a self-assessment form to the user as an expert system. Sites like *trivago.com,* compare the premium rates of all the available hotels and offers the best ones to the user. Although expert systems lack the robust and general intelligence of human beings, they can greatly benefit organizations if their limitations are well understood. It is also said that implementing expert systems in an organization requires a large level of investment, which might sometimes surpass the benefits of using such expert systems. However, considering the benefits such systems provide to an organization, it is often worth implementing. The following table illustrates some of the examples of applications of expert systems.

Finance	Marketing	Human Resource	Manufacturing	Procurement
• Insurance Evaluation	• CRM	• Human Resource Planning	• Production Planning	• Vendor Evaluation
• Credit Analysis	• Market Analysis	• Performance Evaluation	• Quality Management	• Vendor Relationship
• Tax Planning	• Product and Market Planning	• Staff Scheduling	• Product Design	• Equipment Selection
• Financial Analysis		• Pension Management	• Equipment Maintenance and Repair	
• Financial Planning Performance Evaluation		• Legal Advising		

4.9   Relationship of One Information System to Another

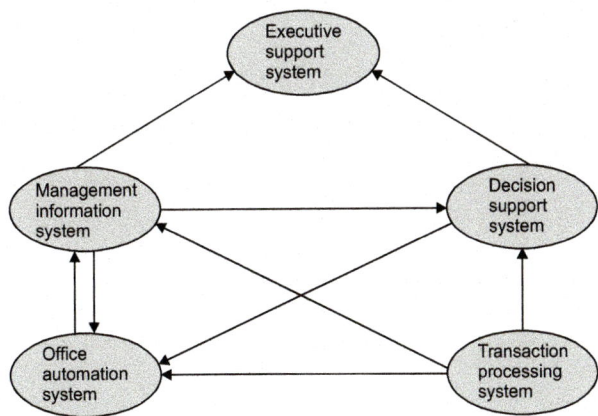

This diagram shows the relationship of various systems working together in an organization.

4.10   Decision Making and MIS

Decision making is selecting the best alternative out of many and involves the following four steps:

(i) Identify the problem

(ii) Diagnose the problem

(iii) Identify the alternatives

(iv) Select the best alternative from which decision will be made

■ **Various Methods That Influence Decisions**

- past experience

- trial and error

- go by the book

- traditional

- logical reasoning

- random choice

4.11   Level of Decision Making

Decision making in an organization can be divided into three categories as described in the following:

■ **Strategic Decision Making:** Decisions taken by the top management of an organization falls under this category. These are the decisions that determine the objectives, policies, and future course of action of the organization, which is a very difficult task. The Board of Directors and CEOs are generally involved in this decision-making process in which the problems are very complex and non-routine.

■ **Management Control Decisions:** Middle-level or tactical-level management is basically responsible for the execution of the decisions made by the top/strategic level of management. They are the ones responsible for the management of the 3 "M"s: man, machine, and material. It is their responsibility to ensure how efficiently and effectively resources are utilized and how well operational units are performing.

■ **Operational Control Decision Making:** Decisions made by the front line or operational-level managers are included in this category. Decisions made at this level determine how to carry out the task set forth by the strategic and middle-level management decision makers. This level keeps track of the performance of the individuals as per set standards and ensures the achievement of the organization's goal.

4.12 MIS Support for Decision Making

Decision making can be considered the single largest activity that is carried out at various levels in business organizations. Herbert Simon's model[1] can be summarized to explain how MIS supports the decision-making process. This model basically consists of three phases as shown in this diagram:

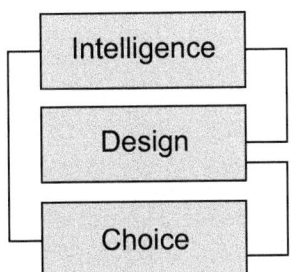

In Simon's model of decision making the first phase is the *intelligence phase*, which deals with the search of situations surrounding organizational environment that demand decisions. Information inputs are accepted during the intelligence phase. The *design phase* is categorized by identifying the various available alternatives, and if any suitable alternative is not available then efforts are made to invent or develop an alternative for the solution of the problem. The viability of the alternatives are evaluated in this phase. The *choice phase* is the one where the optimum alternative is chosen. By optimum we mean that the solution that fits the organization's philosophy and available resources. It is not at all essential that the best alternative is always the best for the organization. In reality, the second-best

[1]www.ccsenet.org/ijbm International Journal of Business and Management Vol. 6, No. 7; July 2011

alternative may be a better choice. The alternative so selected is the decision that is to be implemented.

All the phases of Simon's model depend on MIS to provide relevant data/information for the decision-making process. In the intelligence phase, MIS provides the right data/information at the right time which helps in the processing and evaluation of the data/information accurately. In the design phase MIS helps in the creation and evaluation of alternatives by providing relevant information. MIS even plays a significant role in the choice stage of the decision-making process.

MIS plays a significant role in all types of decisions. The role played by MIS in decision making varies according to the type and the level at which the decision is being considered. It is a well-known fact that humans are greatly affected by the surrounding environment, and this creates numerous stressful situations that negatively affect their decision-making process by blocking the information or misinterpretation of information. In such circumstances MIS acts as a bridge between the environment in which a business operates and the decision maker through information availability and is not at all affected by the environmental factors. Overall, this helps in a better decision-making process.

The information generated by the MIS is considered valuable input for decision-making purposes. For example, the marketing manager uses the sales analysis reports for making decisions such as target setting for a subsequent period, decisions about the performance of a sales person, decisions about product strategies, introduction of new products, and so on. Information generated through the budgetary control system is used to make decisions about approval of expenditure requests. Information generated through the cost system is used to make decisions about pricing. MIS provides a number of tools for supporting the decision-making process such as payoff matrix, decision tree, decision table, game theory applications, and many other tools. Which tool is to be used for the purpose of processing data and information depends on various factors such as nature, type, and the level in the organization where the decision is made (strategic, tactical, or operational). MIS also makes use of all the principals of management, such as optimization technique, simulation, what-if-analysis, trend analysis and projections, cost benefit analysis, HR accounting, and so on.

4.13   Types of Decisions

- **Structured, Programmable Decisions:** These are the decisions that are routine, repetitive, and have a definite handling procedure so they are not treated as "new" every time they are implemented. Because these decisions have a predefined procedure, many can be handled by the operational control personnel. For example *transaction processing systems* (TPS), inventory reorder level formats, rules for sanctioning credits, and so on, are examples of *structured systems*. The information system requirement for structured decisions is:

  - clear and unambiguous procedures for entering the required input data

  - validation procedure to ensure correct and complete input

  - output in a form that is useful for action

- **Unstructured, Nonprogrammable Decisions:** Organizations often find themselves in situations where the decision maker has to use his own judgment, evaluation, and insight when defining the problem. These decisions are strategically important and are non-routine in nature. hence there is no pre-specified decision-making procedure for them. This happens as the situation is totally new or is too changeable to have a stable predefined decision-making procedure. These are the situations that arise suddenly and thus their data requirements are not known in advance. The system should be such that the data retrieval may be possible on *ad hoc* requests. Executive support systems are appropriate information systems support for unstructured decision making.

4.14   Artificial Intelligence Technologies in Business

Artificial Intelligence (AI) technologies are of strategic importance in today's business environment. AI is used in disciplines such as computer science, biology, psychology, linguistics, mathematics, and engineering. Some objectives of AI technology are to develop computers that can think, see, hear, talk, and feel as a human does. It means to develop a system that can use reasoning, can learn from experience, and can perform functions that are normally associated with human intelligence. It is such a buzz word in industry that countries like Saudi Arabia have actually awarded citizenship to a robot named Sofia. It is the first robot in the world to be awarded citizenship. A lot of work is being done to make AI possible, but there are people who are of the opinion that developing technology that makes it possible for a computer to act and think like a human is not possible. But the development in the field of AI is taking place at a great pace and only time will tell if it's possible to build a robot that has human-like abilities. Some of the AI techniques used are defined in the following section.

4.14.1 Neural Networks

A neural network is one of the most popular and widely accepted applications of AI. Its design is based on the brain's mesh-like network of interconnected processing elements called neurons. As in the brain, the network's neurons are connected to all other neurons that are nearby. The interconnected processors in a neural network operate in parallel and interact dynamically with one another. These help the neural network to learn from the data it processes. It is the same process by which the brain learns about all things. For example, if a small child touches a flame for the first time, he is not aware of the consequences. But the resulting burn the child gets remains in his memory; so the next time he does not try to touch the flame. The more data examples the neural network receives as input, the more it learns.

There are two different ways by which a neural network can be implemented on a computer. One is by using software, and the second is by fixing specialized, neural network coprocessor, circuit boards inside of the PC. Apart from this special purpose, neural net microprocessor chips are also available and can be used in specific application areas such as military weapon systems, image processing, and voice recognition. However, most business applications use a neural net software package to accomplish applications ranging from credit risk assessment, signature verification, investment forecasting, data mining, and manufacturing quality control.

4.14.2 Fuzzy Logic

A *fuzzy logic system* is another growing application of AI in business. Fuzzy logic is based on human reasoning which is based on approximation of values and inferences. It is prevalent among humans that they go for approximate values while deciding things. For example, when someone states they are "coming in 5 minutes," it's usually not exactly five minutes, but either less or more. The following example depicts a partial set of rules (fuzzy rules) and a fuzzy SQL query for analyzing and extracting credit risk information on businesses that are going to be evaluated for selection as investments:

Fuzzy Logic Rules
Risk should be acceptable
If debt equity is very high,
then risk is positively increased.
If income is increasing,
then risk is somewhat decreasing.
If cash reserves are low to very low,
then risk is increased greatly.
If PE ratio is good,
then risk is generally decreased.

Fuzzy Logic SQL Query
Select companies
from financials
where
revenues are very large
and PE_ratio is acceptable,
and profits are high to very high,
and (income/employee_tot) is
reasonable.

One can see in this example that words like very high, increasing, somewhat decreased, reasonable, and very low are used in place of crisp data like binary (yes/no). This enables fuzzy systems to process incomplete data and quickly provide approximate, but acceptable, solutions to problems that are difficult for other methods to solve. A few examples of use of the fuzzy logic are washing machines, microwave ovens, autofocus cameras, ACs, etc.

4.14.3 Virtual Reality

Virtual reality (VR) is another important application of AI. *Virtual reality* attempts to enact a human/computer interface that is more natural and realistic. To implement VR the user needs multisensory input/output devices such as a data glove or jumpsuit with fiber optic sensors, tracking handset, video goggles, stereo earphones, and a walker that monitors the movement of the feet. Once all this is in place a user will experience a computer-simulated "virtual world."

VR is very widely used nowadays. Its application areas include medical diagnostics and treatment, scientific experimentation in biological sciences, training of astronauts and pilots, 360-degree product demonstrations on sites, as well as in the entertainment and gaming industry. VR is one application of AI that is getting more widely accepted by the industry, with the only limitation being the cost of its implementation. If one wants high-quality devices, realistic displays, and a more natural sense of motion in VR, it can be expensive.

4.15 Cross-Functional MIS

Earlier businesses were categorized as consisting of various functions, but currently it is said that it consists of a number of processes. Previously, the functions were responsible for their task objectives only, but now all are equally responsible for all the processes taking place in an organization. Likewise, the structure of IS operating in an organization has also changed. Now they are integrated combinations of a number of functional information systems. They support all business processes, such as product development, production, distribution, order and inventory management, customer support and so on. These systems, popularly known as *cross-functional information systems*, cross the boundaries of traditional business functions in order to facilitate the completion of the vital business processes. The cross-functional information systems are used by the organizations as a strategic tool that improves the efficiency and effectiveness of a business process by providing relevant information, and thus allowing a business to achieve its objectives.

Today organizations are using *client–server* technology for IS implementation. This involves installing *Enterprise Resource Planning* (ERP), or Supply Chain management software. The ERP software focuses on the management of all the resources of an organization rather than providing or processing information as per the requirements of specific business functions. One example of this is a new product development process in a manufacturing company. Here it can be seen that a number of departments are involved in the execution of the process. Thus there is a need for a cross-functional information system that crosses the boundaries of several business functions and helps the organization in the successful completion of the process.

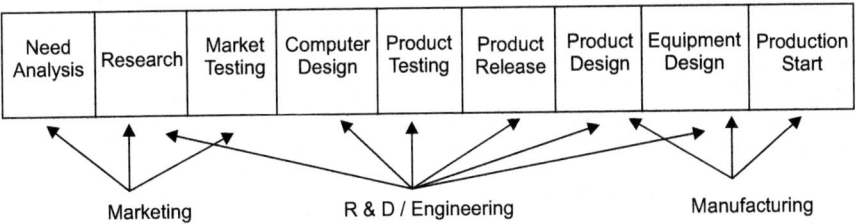

Test Your Knowledge

1. What do you understand about information? How many types of information are there?

2. Discuss the various levels of management existing in an organization.

3. What do you understand about information systems? Explain the structure of IS.

4. What reports can be generated by an information system? Discuss.

5. What do you understand about the quality of information? What are the various characteristics of the information?

6. Discuss the various components of an information system.

7. How many types of information systems are there? Discuss.

8. Discuss all types of transaction processing systems existing in an organization.

9. Describe three of the following systems:
   - Decision support system
   - Executive support system
   - Office information system
   - Management information system

10. Draw a diagram showing the relationship of various information systems with one another.

11. MIS is an evolving concept. Discuss.

12. What do you understand about management information systems? Discuss.

13. Discuss why there is a need for information. On the basis of need, in how many categories can information be divided.

14. What do you understand about the value of information? Discuss.

15. What do you understand about decision making? Explain the process of decision making.

16. What are the various levels of decision making in an organization? Discuss.

17. How is MIS useful in the process of decision making? Discuss.

18. Discuss Simon's model of decision making.

19. Discuss the various levels of information required by the different levels of management in performing the management functions of marketing, production, finance, human resources, etc.

20. Explain the difference between structured programmable decisions and unstructured nonprogrammable decisions.

21. What do you understand about cross-functional MIS? Discuss.

22. What are the various artificial intelligence technologies that are getting used in business? Explain them.

23. Explain the following:

- Neural network

- Fuzzy logic

- Virtual reality

24. What do you understand about an expert support system? Explain.

25. Explain the following:

- Marketing information system

- Finance and accounting information system

- Computer-based personnel information system

CHAPTER 5

APPLICATION OF CROSS-FUNCTIONAL MANAGEMENT INFORMATION SYSTEMS

5.1 Enterprise Resource Planning

Enterprise resource planning (ERP) is a buzz word in the industry. ERP not only includes the application of information technology, but it also incorporates corporate mission, attitude, belief, values, objectives, style of operation, and people who make the organization. ERP can be defined as a system that combines all the aspects of a business. It takes care of all activities within the organization including planning, staffing, organizing, controlling, manufacturing, marketing and sales, etc.

Today an organization can sustain only if it has the ability to respond to changing customer needs and also has the ability to grab the opportunities in the market when they arise. It is a fact that only those organizations can survive in the market that are continuously striving to improve their performance in the areas of quality, customer satisfaction, performance, and profitability. This is a 24x7 job and is beyond human capabilities. Thus a need arises for a system that can provide information across all functions and locations within the organization. The ERP is a system that has the capabilities of assessing the resources, required for a given situation for achieving related business objectives, It also helps in the execution of strategies, plans, and actions in a timely manner.

The ERP is a system that provides support for transaction processing, required modification, and reporting across business functions.

ERP as a package covers all major functions of a business, creates a database, and also helps in the compilation of a knowledge base. This helps in the planning and control of a business. A typical ERP package solution has the following modules:

- marketing and sales, marketing
- manufacturing
- finance
- personnel
- maintenance
- purchase, distribution, and inventory management
- planning and control

These modules perform all the necessary activities such as capturing data, transaction processing, data validation, what if analysis, report generation, etc. The ERP usage is controlled at all levels by providing authorization. ERP provides a system of check and validation of the users and so maintains the integrity of the system. The user is only allowed to proceed after the authentication process takes place. For example, the front-line managers are allowed to pass cash transactions of 5 lakhs only if the transaction is above 5 lakhs, would they not be able to execute it.

5.2   Objectives of ERP

There are several advantages to having an ERP system in place. Some of them are that it:

- provides the support required for the completion of the business processes
- improves the productivity in the implementation of the processes
- gives customer an option to modify the business process as per their needs

For example, Airtel gives it customers the option to fix the mobile plan for themselves by setting the calls, number of SMSs, and Internet limit for their post-paid plan.

5.3   Benefits of The ERP System

There are numerous benefits of implementing an ERP system in an organization; a few of which are summarized as follows:

- It helps in the optimum utilization of resources
- As all functions are interconnected, it results in an increase in productivity
- Since the entire organization is available online, it increase the operational transparency and customer satisfaction
- Transparency
- Delegation of decision making
- User friendliness
- Faster processes
- Improvement in the quality of decision making
- Instant inducement through updates

5.4   ERP Solution Structure

The Enterprise Resource Planning solution structure is built in three layers namely technology, business, and implementation.

- **Technology:** The technology used in developing an ERP solution is managed through the database management system that helps from data acquisition to database creation, updates, and maintenance. The application development is done through the client-server technology, where the server handles the specific or the general functions as the case may be, and the client plays the role of processing interactively and locally for meeting the information needs.

- **Business operations:** On the business side, ERP provides solutions for data entry, data capture, and transaction processing and database update.

- **Implementation:** The ERP implementation is multi-user and calls for the network usage for the work flow, communication, and the access to the database, which may be at one location or distributed.

The successful implementation of the ERP calls for a strong technology component appropriate to the environment.

5.5   Basic ERP Features

Features provided by an ERP system consist of all the options required for the fulfilment of all the functions to be performed by an organization. Some of those functions are help functions, action messages, MIS for strategy monitoring and control, order processing, receivable analysis, sales forecasting and budgeting, personnel management, work in progress tracking, inventory analysis and management, material receipt and issue system, purchasing and procurement, assets accounting, working capital management, cost benefit analysis, human resources management, planning, recruitment, training, and so on.

5.6   ERP Selection

The selection of ERP can be made on the basis of following three dimensions:

5.6.1 Vendor Evaluation

It is based on:

R & D investment in the product

Future plans of the vendor

Business philosophy of the vendor

Ability to execute the ERP solution

Image in the business and in the information technology world.

5.6.2 Technology Evaluation

It is based on following factors:

Hardware-Software configuration

Operating system and its level of usage in the system operations.

Support system technologies like bar coding, EDI, imaging, communication network, etc.

5.6.3 ERP Solution Evaluation

It is based on following factors:

- Solution architecture and technology

- Product rating in its class of products

- Flexible design

- The ability of a quick start on implementation

- Ease of use, ease of learning, implementation and training

An ERP system is a tool that helps in the management of the enterprise resources. One has to remember that ERP is only a supporting system and is not a magic wand that solves all the problems of an organization. An implementation of the ERP system requires a total commitment and full participation of the organization. It is not a labor-saving device, rather it's to be considered a managerial tool. Since it is designed for the rise of productivity, the management must exploit it to its advantage by adopting the best practices or changing the practices through business process re-engineering.

ERP implementation follows the waterfall mode approach, which means the progress of system development progresses largely in one direction. Once an organization decides to develop an ERP system and the vendor is selected for the development, the implementation begins with first meeting between the vendor and the management. In this meeting the vendor and the management discuss the organization vision, mission, and major implementation issues. Implementation of ERP in an organization is a long-term activity, so preliminary planning is done with the requirement analysis and then gradually proceeds further.

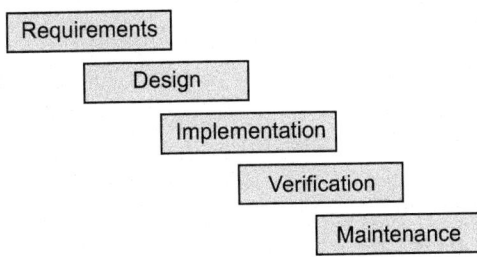

5.7　Customer Relationship Management

Customer relationship management (CRM) is a concept that has been coined by the industry and has evolved from sales automation. Sales automation evolved into customer assets management and then into CRM. CRM means having a complete focus on the customer with an objective of building a long-term relationship. Peter Druker, in his book *Marketing Management*, suggested that creating a new customer is five time costlier than retaining an old customer. CRM works on the same philosophy, that is, to always maintain a high-level relationship with a customer so it becomes possible to retain them. If organizations want to succeed the path is clear—companies need to exceed their customers' expectations at all times, they need to pamper their customers. Organizations believe that this goal can be achieved if they put their basic business activities such as marketing, sales, and financial transactions on the Web and let the customer serve themselves. This process allows customers to feel as if they are an extension of the organization. For example, the huge success of the Indian Railway Catering and Tourism Corporation site that allows the Indian passengers to book, cancel the railway ticket on their own via net (where as this is the job of a booking window clerk).

In a cut-throat environment, when all promotional channels are saturated with the product awareness messages, there is a need for marketing programs to be focused on value-oriented differentiation. Hence today businesses are reaching out to say "We are here for you," rather than what they used to say, simply "We are here." This process was first started by ICICI in its promotional advertisements when it started its banking services in India.

Customer turnover is one of the biggest problems faced by the organizations today. All the marketing efforts made by the organizations are failing apart. Customer loyalty is hard to find, even though the marketing efforts may be attracting new customers. To add to it, every organization has a specific segment of a customer and they look for the loyal customer in that segment. Studies have shown that the customers with whom the business has constructively established a relationship are the customers who bring their business back. Thus one can say that the logic behind focusing on CRM is to have a single enterprise view of the customer with an objective of developing a long-lasting relationship, which in turn will lead to improved profits. This can be done only by identifying all the products, services, and intermediary relationships that a customer has with the organization, as well as having a history of all the interactions that have taken place between the customer and the organization since their relationship started.

5.7.1 CRM and e-CRM

CRM consists of a person-to-person interaction with the customer, whereas in e-CRM the interaction is done electronically via website, e-mails etc. e-CRM enables an organization to make available its functions and processes to all stakeholders, especially customers in a manner such that it offers new opportunity to know one another better, add value to the processes, gain new economies, and reach new customers which was otherwise not possible with face-to-face CRM practices. With e-CRM it is possible for an organization to remain in contact with its customers 24x7. By providing the FAQ section on the site or offering online chatting facilities it is becoming possible for organizations to pacify the customer round the clock.

5.8 Security and Ethical Challenges of It

With the deepest possible penetration of information technology (IT) in the urban population, major challenges are emerging. Maintaining privacy, security of data and identity, copyright violation, and numerous computer crimes are few of them. There is many in society who are using technology for a good cause, but there are also people with negative orientation who eagerly await to utilize the loop holes that are often found in technology. Cybercrimes are at an all-time high, and many individuals and organizations are at risk of becoming cyber victims on a daily basis. Few of the challenges faced by IT are described in the following sections.

Security Challenges of IT

- It is very simple for a hacker to hack into any computer or system as it gets connected to the internet. Using an IP address, a hacker can easily access a computer and can transfer data.

- Cookies are another problem that most people do not understand. Cookies are small programs that collect information whenever a user is logged in on the Internet. Nowadays, a number of companies use cookies to determine which products or service they can advertise to particular users, thus posing a high risk of fraud and conflicting interests.

- Online banking is an area of major concern for security. The transfer of money can easily be redirected by a hacker to any other accounts and affects both the bank and the customers who are using online banking technology.

- Webcams pose a big threat toward the privacy of a computer user. It is very easy to turn on any webcam of any computer online and then gain access to one's private life.

- Social networking sites like Facebook and Instagram offer users an option to find and connect with new and old friends, as well as share information, videos, and photographs. Even the live-streaming options are now provided by these sites. It is however very risky as users share their personal information, and individual or family photos on these sites because they are easily accessible even to unknown users. It is quite possible that this may expose someone to somebody who has bad intentions and may use the data in an unwanted way. Also, some companies are known for spying on their employees via these social networks.

Ethical Challenges of IT: Major ethical issues are defined in the following sections.

Copyright Infringement

- Information technology has made it easy for users to access information and use it without proper reference or permission.

- There are numbers of sites that allow users to download free old or new music, movies and books, thus making the original creators of these works lose credibility of their works, because now it is very easy to gain access to such material and share it. It also happens that a single user can buy the original work and then upload it to a site from where others can download it for free.

- Software is easily available and can even read and download protected data. This is a big ethical issue faced by users of IT.

Increased Pressure on Software Experts

- Most organizations are now available online 24x7. This puts a lot of pressure on people working in an IT department since it is their responsibility to ensure the flawless working of the systems. This creates stress and work overload which results in poor performance and mistakes.

Digital Divide

- IT offers numerous opportunities and has changed the way businesses operate. However, it also brings with it certain problems. Organizations need to train their workforce to adopt the new culture although it is a bit costly.

Employment Challenges: The impact of information technology on employment is a major ethical concern, and those who cannot keep pace with the technological advancements are likely to fall out of the race and thus will lose their job.

5.9   Business Ethics

Business ethics refers the ethical and moral issues that a manager faces while performing their routine work. Business ethics means following the principles and adhering to the norms that are sometimes very difficult to follow in a tough business environment. Business ethics are subdivided into two groups. The first deals with the illegal, unethical, or questionable practices of organizations, and the second deals with the ethical questions that managers are confronted with during their daily business decision-making process. Managers are sometimes forced to deal with the dilemma of *what to do* and *what not to do* because their prime concern is to generate revenue for the organization. Sometimes, if they handle the issue ethically they might not be able to achieve their goals. Thus a question that arises is "How can a manager make ethical decisions when confronted with business issues?" Several important alternatives based on theories of corporate social responsibilities can be used.

- **Stockholder Theory:** This theory states that managers are the representative of the stockholders, and their only ethical responsibility is to increase the profit of the business without breaking the law or involving in any malpractices.

- **Social Contract Theory:** This theory is considered as the best practice, in that it states organizations are ethically responsible to all the members of society in which they operate. This theory requires organizations to increase the economic satisfaction of consumers and employees by providing them reasonably priced products, reasonable salaries, treating employees decently, avoiding malpractices, and avoiding practices that systematically worsen the position of any group in society. Apart from it, environmental concerns have a prominent place in this theory.

- **Stakeholder Theory:** This theory suggests that a manager has an ethical responsibility only for the benefit of the stakeholders of an organization. This usually includes the corporation's stockholders, employees, customers, suppliers, and the local community.

5.10   Cybercrimes

Cybercrime is a growing threat to society caused by the criminal or irresponsible actions of individuals who are ready to take advantage of the ignorance of computer users, as well as the susceptibility of computers and the internet. They are emerging as a major challenge for the computer world. A few of the computer crimes are summarized as follows.

1. **Computer Crime:**
   (a) The unlawful use, access to hardware, software, data, or network resources.
   (b) The illegal software copying or release of personal information.
   (c) Blocking a user's access to their own hardware, software, data, or network resources.

2. **Hacking:** Hacking can be defined as a fanatical use of computers to gain unauthorized access to someone else's network or computer system. Unethical hacking is causing a danger to computer users. Fanatical computer users called hackers alway try to break into other networks to steal or damage data and programs of other users. Hackers are capable of monitoring e-mails, file transfers, extract passwords, or steal files/data, and cause irreparable losses to the common users of the system.

3. **Cyber Theft:** It means theft of money. In most of the cases it is an insider job that has the access to the network or the passwords.

4. **Unauthorized Use at Work:** Utilizing an organization's computer, software, or network during the office hours for personal use is called time and resource theft. For example, it is quite common for employees to log on to Facebook while doing their routine work. Organizations uses intranet to avoid this unauthorized use of resources.

5. **Software Piracy:** Not buying the legal copy of software is a big problem in the IT industry. Online tools are available to help users download pirated copies of software from the Internet.

6. **Computer Virus:** Computer viruses are small programs that get copied to a computer system and attach themselves to the programs while they continue multiplying, and thus gradually reducing the speed of the system and corrupting the executable files. It is the most destructive example of computer crime.

7. **E-mail related crimes:**
   - **Email spoofing:** Refers to receiving mail that seems to generate from one source, when it has actually originated from another source.

   - **Email spamming:** Refers to sending unsolicited email to thousands of users while concealing the identity of the sender.

   - **Sending malicious codes through email:** Malicious codes or links to websites are sent as e-mail attachments. As soon as the users click on attachment or the link, the virus gets copied onto the hard disk of the system.

   - **Email bombing**: Refers to the sending the same message to the same recipient again and again.

   - **Sending threatening emails**

   - **Defamatory emails**

   - **Email frauds**

8. **Cyber Terrorism:** Terrorist organizations are using the internet as their new tool for spreading hatred and the recruitment of misguided youth by using internet telephone services very effectively. Cyber Terrorism is becoming a favorite an option for terrorists for reasons cited below:
   - It is difficult to trace, secure, and cheaper than traditional terrorist methods.

- They can easily target any group within any society.

- They can operate from locations that are difficult to locate.

- It has a cascading effect on people across the globe.

9. **Banking/Credit Card-Related Crimes:** Online banking is an area that is greatly affected by cybercrimes. More and more users are opting for online banking, but sometimes from an unsecure system that runs on pirated software without installing an antivirus, thus making them vulnerable to bank fraud. The use of stolen card information or fake credit/debit cards are common.

10. **E-commerce Fraud:**
    Sales and Investment fraud:

 ▪ No one controls the Internet so anyone can upload any information they choose. As such, there are numerous incidents of fraud cases happening online. Countless offers appear on any given number of sites luring users into investments or loans that claim to have very high returns. By the time users realize it's a trap, they lose their money.

 ▪ Users order from an online site thinking they are buying the desired product, pay money, but never get the delivery of the product, or are sent substandard products.

 ▪ Products that are otherwise banned in the public domain are easily available online. Users can be lured to buy narcotics, prohibited medicines, weapons, endangered species of wildlife, etc., simply by posting information on websites, bulletin boards, or simply by posting a message.

11. **Online gambling:** Numerous websites offer online gambling. It is assumed that most of these sites are actually fronts for money laundering.

12. **Defamation:** Defamation means writing something degrading about someone or posting obscene photos of another person. For example, publishing derogatory matter about someone on a website or sending e-mails containing derogatory information to the people who are known to that person. It is also known as cyber smearing.

13. **Cyber Stacking:** Cyber stacking means following user's movements across the net by posting messages on sites or chat rooms visited by the user, and also e-mail booming of the victim, etc. The stacker usually does this to cause emotional distress to the victim.

14. **Identity Theft:** It is an emerging field of cybercrime, especially in United States. Identity theft is when someone downloads identity data of someone else without their knowledge and then commit theft or fraud using that other person's identity.

15. **Data Diddling:** Data diddling is when users enter some data in a system, but other data gets entered. It is either a virus that changes data, or the programmer of the database or application, or anyone else who was involved in the process of developing the software.

16. **Theft of Internet Hours:** This means someone pays for internet hours and someone else uses them. By gaining access to an organization's telephone switchboard, individuals or criminal organizations can obtain access to dial-in/dial-out circuits and then make their own calls or sell call time to third parties.

Privacy Issues

It is very easy as well as viable to assemble or disseminate information by the use of information technology tools. This feature of IT also raises some privacy issues that are defined in the following:

- **Violation of Privacy:** It is possible to access anyone's private e-mail conversations and computer records by cracking the password.

- **Computer Monitoring:** Refers to tracking the movement of a person. It's not uncommon because has a mobile phone and spends a lot of time online.

- **Computer Matching:** Cookies are used to get customer information and unsolicited mails are sent to them for marketing additional business services.

- **Unauthorized Personal Files:** Collecting telephone numbers, e-mail addresses, credit card numbers, and other personal information to build individual customer profiles.

Test Your Knowledge

1. What do you understand about enterprise resource planning (ERP)? Explain with an example.

2. What are the various objectives of an ERP system? Discuss its various advantages.

3. Discuss the ERP solution structure.

4. What are the basic ERP features? Discuss.

5. What criteria is followed while making an ERP selection?

6. What do you understand about CRM? What is its relevance in today's business environment?

7. What is e-CRM? Discuss.

8. What are the various security and ethical issues of IT? Discuss.

9. What do you understand about business ethics? Explain with an example.

10. Describe cybercrime. What are the various categories in which cyber-crime can be classified?

11. What are some various privacy issues? Discuss with an example.

INDEX